Manufacturer's Guide to

Implementing the Theory of Constraints

Manufacturer's Guide to
Implementing the Theory of Constraints

Mark J. Woeppel

CRC Press
Taylor & Francis Group
Boca Raton   London   New York

CRC Press is an imprint of the
Taylor & Francis Group, an **informa** business

Library of Congress Cataloging-in-Publication Data

Woeppel, Mark.
  The manufacturer's guide to implementing the theory of constraints /
Mark Woeppel.
      p. cm.
  Includes bibliographical references and index.
      ISBN 1-57444-268-6
      1. Production management. 2. Theory of constraints (Management) 3.
Reengineering (Management) 4. Quality control. I. Title.
      TS155 .W58 2000
      658.5—dc21

                                                                    00-011198
                                                                    CIP

© 2001 by CRC Press

No claim to original U.S. Government works
International Standard Book Number 1-57444-268-6
Library of Congress Card Number 00-011198
Printed in the United States of America   6 7 8 9 0
Printed on acid-free paper

Table of Contents

Introduction

This book is a guide and workbook for anyone wanting to implement Constraint Management in a manufacturing organization.

I will show you a *process* to implement that has been proven in multiple organizations and will provide you with short cuts to a successful implementation. You'll find a wealth of resource material in the boilerplate procedures and policies. I have deliberately emphasized the tactical processes and left the strategic area to future writing. Most implementations are less than 5 years old, so little has been done in this arena. There are implementation checklists with sample policies and procedures. You'll get an understanding of why these policies are there — and their importance to the success of your implementation. My intention is to focus on the practical aspects of the implementation, rather than the theoretical.

How Do You Do This Stuff?

Quite a bit has been written about constraint management and the theory of constraints. Many people have written their success stories. There is a lot about how it works, but not how to do it, how to answer the question, "What in my business has to change in order to be successful?"

The content is based on my implementing experiences and describes how to make the theory practical. I hope, that by sharing my experiences, you will avoid some of the mistakes my clients (and I) made. I'll take you from the first day you decide you want to change your organization, through the writing of procedures, to the "anchoring" of the implementation, into the cultural practices of the organization.

What Is a "Successful" CM Implementation?

I don't judge success against an indicator of x percent improvement of profit or return on investment. If you use the concepts, you will improve performance. No doubt about it. I have NEVER seen an implementation where the concepts were applied correctly and the organization didn't see significant bottom line results.

A successful implementation is one where the management team is, on a regular basis, considering constraint implications in the daily decisions of allocating resources and making customer commitments. The management team is making the long-term operations decisions (outsourcing and resource acquisition) using similar criteria. So, focus on the cause of the success, not the effect.

How This Book Is Organized

I've organized the book from basic concepts to more complex concepts. If you look at the table of contents you'll notice that the first thing that I do is explain what constraint management is. My intention is not to give a detailed explanation, but an overview. I assume you know something about constraint management and you are interested in the topic. In the event that you do not, I hope to give you enough detail that you will be able to make some sense of the system that is going to be implemented and you'll understand what I'm talking about.

The first part of the book gives an overview of constraint management and an overview of the implementation process. This is followed by a description of the system being implemented. My thinking is that by my giving a bigger picture first and then later going into more detail, you will be able to more readily digest the material. You can be confident that if I'm mentioning something, it will be explained in more detail later.

Following this, you'll see a discussion of the implementation process where I give a summary of each step required to implement the constraint management system. The subsequent chapters break the implementation process down into sections. Within each section, you will see detail on what the procedure is and the steps required to complete it.

Acknowledgments

I don't think anyone writes a technical book that is truly original; all build on prior knowledge. This book is no exception. Each of my employers and clients had a hand in "inventing" the policies and processes I've described. I am very grateful for the experience of working with them. Many customers became my friends.

I want to thank Eli Goldratt, who has had the single largest influence on how I think about business and who I have the privilege of calling my friend. He has literally changed the course of my professional life. Robert Fox taught me about being a consultant — I was very wet behind the ears and he took a personal interest in my development as an associate of the Goldratt Institute. Thanks. Bob LaCourciere was my first client. He believed in me and allowed me to experiment. We were a very good team. Bob Page was another client who helped me a great deal. Much of the development of the policies and procedures took place on his dime.

The book was a team effort. Although my name appears on the front, Scott Monaco did a lot of the artwork (the good ones). Lisa Scheinkopf helped me do a better job of explaining myself. It hurt, but I think the book is better because she gave me some of her time. Linda Poling also helped with editing. Will Baccich helped me see where additional emphasis was needed. I'm proud to call you my friends. Thanks for your help.

Mark Woeppel
Carrollton, Texas
July 3, 2000

The Author

Mark Woeppel, CPIM, is a master of positive transformation of manufacturing organizations, with over 10 successful turnarounds. He is expert in materials management and synchronous flow manufacturing, with an emphasis on implementing high performance business systems in manufacturing. Dr. Eliyahu Goldratt, author of *The Goal*, calls him the most skilled implementer in North America.

Mr. Woeppel is a nationally known speaker and lecturer on the topics of continuous improvement, synchronous flow manufacturing, cost accounting, organizational measurement, and functional alignment and integration of organizations. Notable venues are: Northwestern University Kellogg School of Business, University of California, Los Angeles, American Production & Inventory Control Society, and Institute for Management Accounting.

His industry experience includes the graphic arts, industrial equipment, consumer products, automotive, oilfield equipment, electronics, and steel fabrication.

He serves as President for Pinnacle Manufacturing Consulting, provider of consulting, training, and technology solutions in the implementation of supply chain synchronization and specializing in the implementation of drum–buffer–rope scheduling systems, theory of constraints, and advanced planning and scheduling (APS) software.

About APICS

APICS, The Educational Society for Resource Management, is an international, not-for-profit organization offering a full range of programs and materials focusing on individual and organizational education, standards of excellence, and integrated resource management topics. These resources, developed under the direction of integrated resource management experts, are available at local, regional, and national levels. Since 1957, hundreds of thousands of professionals have relied on APICS as a source for educational products and services.

- **APICS Certification Programs**—APICS offers two internationally recognized certification programs, Certified in Production and Inventory Management (CPIM) and Certified in Integrated Resource Management (CIRM), known around the world as standards of professional competence in business and manufacturing.
- *APICS Educational Materials Catalog*—This catalog contains books, courseware, proceedings, reprints, training materials, and videos developed by industry experts and available to members at a discount.
- *APICS—The Performance Advantage*—This monthly, four-color magazine addresses the educational and resource management needs of manufacturing professionals.
- *APICS Business Outlook Index*—Designed to take economic analysis a step beyond current surveys, the index is a monthly manufacturing-based survey report based on confidential production, sales, and inventory data from APICS-related companies.
- **Chapters**—APICS' more than 270 chapters provide leadership, learning, and networking opportunities at the local level.

- **Educational Opportunities**—Held around the country, APICS' International Conference and Exhibition, workshops, and symposia offer you numerous opportunities to learn from your peers and management experts.
- **Employment Referral Program**—A cost-effective way to reach a targeted network of resource management professionals, this program pairs qualified job candidates with interested companies.
- **SIGs**—These member groups develop specialized educational programs and resources for seven specific industry and interest areas.
- **Web Site**—The APICS Web site at http://www.apics.org enables you to explore the wide range of information available on APICS' membership, certification, and educational offerings.
- **Member Services**—Members enjoy a dedicated inquiry service, insurance, a retirement plan, and more.

For more information on APICS programs, services, or membership, call APICS Customer Service at (800) 444-2742 or (703) 354-8851 or visit http://www.apics.org on the World Wide Web.

1 What Is Constraint Management?

In his book, *The E Myth*, Michael Gerber discusses the entrepreneurial myth and the evolvement of businesses. He outlines two distinct approaches to building an organization, what I call a systems approach and a craftsman approach.

Craftsman Approach

The defining element of the craftsman approach is that it relies heavily on experts (craftsmen) to accomplish the organizational objectives. In this organization, you will find that:

1. Managers are "doers." The leaders of the organization are also the people who make things happen in the company. They may wear multiple hats. The line between those who make the product and those who manage the process by which the product is made is blurred.
2. Leaders have high personal involvement in the order fulfillment process. The chief expediter is usually the president or the general manager. Almost no order ships on time unless someone gives it special attention.
3. There is heavy reliance on "key" people to accomplish the daily activity of satisfying the customer. These people are not necessarily management, but generally are.
4. There are long training periods for every leadership position — almost an apprenticeship. Because there is no system, the leader/manager must be in the position a long time before he or she "learns" it. It takes time before the majority of situations can arise and someone can learn the appropriate responses.

An example of a business using the craftsman approach would be your corner restaurant. It is owned by a husband and wife, and the food is pretty good — excellent, actually. But when the owners are gone, something is not the same. Maybe you do not receive the same friendly, speedy service. Perhaps it is not as clean as it usually is. You can tell the craftsman is gone.

The consequence of this approach is that management spends more time working *in* the business than *on* the business. So, when the management experts are gone, things do not run well.

The main idea of the craftsman approach is that management expertise is gained primarily through experience. Management of the business relies on internal experts and their "art" or judgment. Decisions are not quantified; the primary reliance is on the crucial managers.

The Systems Approach

The systems approach is not as dependent on the experts. Management strives to create a repeatable process for managing the organization — a process that can be taught to others. Managers try to create a procedural model of the business, as if they had to duplicate the business as a franchise.[1]

You will see the following attributes in a systems-oriented organization:

■ Systems handle the day-to-day work of delivering customer outcomes.
■ The business runs mostly on "auto pilot" — without direct management intervention.
■ The effort of managers is almost exclusively on future planning.
■ Training is focused on teaching the system.

The main idea is that the system runs the business, the people are taught to run the system.

An example of a systems approach business is McDonalds™ restaurants. No matter your opinion of the food, you consistently receive a quality product and reasonable cost, quickly. You also know that McDonalds is one the most financially successful firms in history. It has achieved the goal of the systems approach — consistency and reliability.

---

[1] The franchising concept works because it employs a systematic approach to running the day-to-day aspects of the business. The franchisee, when he purchases a franchise, purchases not just the name, but the policies and processes to run a successful business. This is what gives franchise businesses significantly higher success rates than individual start-up businesses.

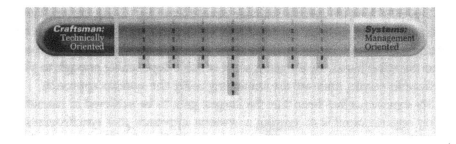

Figure 1.1   The Craftsman–Systems Continuum

No organization takes a purely systems or purely craftsman approach. Even the most systematic organizations will have some technical orientation; even the most technically oriented organizations use some systems (see Figure 1.1). Your organization will fall somewhere in between these two extremes — perhaps, since you are reading this book, your organization is more on the craftsman side.

So why is it important which approach you use? For one thing, a systems approach delivers consistent results. The results I'm talking about are those that the customer values — reliability, responsiveness, and consistency. I believe the most significant contribution total quality management (TQM) has made to manufacturing is that it approaches the process of manufacturing a product as a system that can be controlled and improved. The idea of improving product quality by improving the system that delivers the product was a breakthrough in management thinking. It is contrasted with the craftsman approach, which by its nature cannot deliver a consistent result over time. It cannot because it is not designed to do so. The craftsman approach is focused on the people delivering the results of the system, not the people receiving the results. The decision-making processes — those processes that are geared to delivering product to customers and results to stockholders — are systems.

Why Cannot Manufacturing Employ a Systems Approach?

Some people contend that running a production facility effectively has more to do with competent management (craftsmen) than any other factor. Some go as far as calling production management an art.

Typically, how do people manage manufacturing? Using the craftsman approach or the systems approach? If we depend on the experience and intuition of certain individuals, then we are using the craftsman approach. If, on the other hand, we are using a set of rules dependable enough to yield predictable results, then we are managing using the systems approach.

My experience has been that the largest gain to be achieved in manufacturing organizations is to employ a systems approach that synchronizes or aligns the internal functions of the enterprise. If these functions are synchronized, productivity improvements of 15–20% will be seen the first year. Cycle time will drop by 50–75% with a favorable bottom-line impact.

FMC Wellhead Equipment Division is located in Houston, Texas and provides high performance wellhead drilling and completion equipment for land and offshore oil platform applications. After they adopted the systems approach, inventory was reduced 50%, saving over $2.8 million per year. With the found excess capacity, some outsourced business was brought back in-house lowering expenses by $400,000 per month. They continue to earn significant profits.

Dixie Iron Works, in Alice, Texas improved profits 300% and grew revenues 50%.

So — why cannot manufacturing employ a systems approach? No reason, they can and should.

Developing a Systems Approach — Constraint Management

Dr. Eliyahu M. Goldratt is widely considered the inventor of constraint management. First explained in his best-selling book, *The Goal*, the concepts were first articulated in the late 1970s. The techniques have been called Synchronous Manufacturing, OPT® Principles, Commonsense Manufacturing, Continuous Flow Manufacturing, and other terms. For the purposes of this book, my working definition of constraint mangement is "a philosophy and set of techniques to manage and optimize[2] the activity of the business."

Implementing constraint management (CM) is the quickest way to implement a systems approach. However, the systems approach I am talking about is mainly focused on the decision making and priority setting of manage-

---

[2] Some will argue that optimization is pointless. One should only do better than one did yesterday, and, in principle, I agree. However, when I say optimization, I am referring to a methodology to make the best decisions possible, given the amount of information available and using the best judgment of managers.

ment, the most important systems in the organization. There are some specific procedures you can employ, but CM provides a framework for making decisions to implement appropriate procedures, which leads to a systems approach.

I cannot begin to provide a comprehensive description of CM and there are already excellent texts that do just that. What I hope to do is give you a flavor of the significant issues of a CM implementation.

Measurements

A major impediment to using a systems approach is alignment of purpose and behavior. In order to implement, we have to bridge the gap between day-to-day decision making (and behavior) and the purpose of the organization. The measurement system is the key to accomplishing that.

The measurement system must be simple and easily understood by those being measured. Make it too sophisticated, and it will be difficult to sift the important from the unimportant. It must be useful for the day-to-day decision making in the company. The system should direct managers to focus on the specific outcomes the organization wants to see.

The measurements provide the basis for a common language in the organization and communicate the priorities — the "rules" of the company. What is measured communicates what is important. As an analogy, the game of football cannot be understood without the scoreboard. Tactics are dictated by the current down, how many yards to the first down, how much time is left on the clock, and so forth. You cannot expect to win if the team does not know (a) the game, (b) the rules, and (c) the score. Every day, your people show up and play a game called "Let's make money." Unfortunately, most people in the company do not know that game, so they play "Let's collect a paycheck." Quite a different game. Having the proper measurements will not make you an excellent team, but you cannot have excellence unless everyone understands what is important. The measurement system accomplishes that end.

Measurements also (indirectly) create behavior. An axiom of management is, "What is measured improves." How to explain this fact? Measures always affect people. Whether it is the activity or the results you are measuring, there is a *person* being measured. If we accept that people want to "win," we give them the "score," and we encourage the behaviors to improve the score, we will see an improvement in the measure.

The foundation of a good measurement system begins with the articulation of the goal or mission of the entity being measured. Since we are dealing with for-profit entities, we can use Dr. Goldratt's definition for them: To

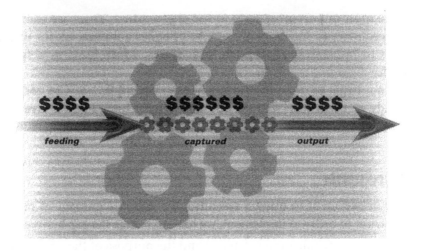

Figure 1.2 Money Machine

make more money, now and in the future. Therefore, the measurements will be based on the ability to make money. This ability is measured in absolute terms as net profit, and in relative terms, as return on investment.

A Money-Making Machine

The difficulty in defining a measurement system comes when we attempt to tie local actions to these two measurements. As stated earlier, most people don't realize that this is how the score is kept in their company. If they do understand it, there are few tools to guide their decision making — to quantify local decisions based on the global financial measures.

Before I resolve the difficulty, I need to develop the concept further. I will use the analogy of the company as a machine. This machine produces money, as does every other for-profit company in the world. In this machine, there is money being produced (I hope), money captured, and money entering (Figure 2.1).

Throughput

The money produced by the system is called *throughput*. Throughput is "all the money generated by the system." Some refer to this as the *throughput value-add*. It is not production or parts related; a part becomes throughput when the customer is invoiced.

Throughput is not gross revenue. Some revenue collected by our system is not throughput produced by us, but by our vendors. This part of the revenue merely passes through our system. Therefore, throughput is calculated by taking gross revenue minus all totally variable (with the order shipped) expenses — purchased material cost, sales commissions, any subcontract expenses (again, associated with the orders being shipped), and freight if you prepay and then add.

Inventory

The money captured within the system is called *inventory*. Typically, we think of inventory as parts, but since the product of the system is money, the inventory is money. Inventory is defined as "all the money the system invests, purchasing items it intends to resell."

All assets (buildings, equipment, fixtures, etc.) are inventory (at depreciated cost). All parts, at purchase cost, are inventory. The concept of *inventory value-add* (labor and overhead allocation) is not used in constraint management. One adds value to the system, not the product of the system. Products do not really have a value until someone purchases them.

Additionally, "inventory value-add" distorts the decision process, forcing management to rely on intuition and manipulation of the numbers to arrive at a good decision. To understand further how the inventory valuation concept distorts the decision process, I suggest that you read Chapter 2 of *Synchronous Management* (Srikanth and Umble, 1997).

Operating Expense

All the money spent to turn inventory into throughput is *operating expense* — all direct and indirect payroll expense, all supplies, all overhead. All expenses related to time are operating expense. In general, operating expense is the real money you take from your pockets to produce products or services to satisfy the customer.

Since labor is related to time, it belongs under operating expense. This concept stands in direct contradiction of generally accepted accounting practices. I am not trying to make an argument against GAAP, rather in favor of good decision making. Labor expense is a function of time in the system, not product shipped. The system's payroll expense doesn't change one bit whether you make 50 or 500, whether people make good parts or bad parts. In most cases, putting labor into operating expense simplifies the decision process and makes measurement much easier to understand.

Figure 1.3 Categories of Money

Global Measurements

Throughput, inventory, and operating expense (T, I, OE) can be tied to the global measures of the system: net profit and return on investment (Figure 1.3). Net Profit equals Throughput minus Operating Expense (NP = T – OE) and Return on investment equals Net Profit divided by Inventory (R = NP/I). Productivity is Net Profit divided by Operating Expense (P = NP/OE). Inventory Turns are Throughput divided by Inventory (IT = T/I).

Measures in Practice

Suppose I want to purchase some capital equipment; perhaps I will buy a new machine or improve a tool. The standard calculation to justify my decision is illustrated in Figure 1.4. Although this is the "typical" justification, there are three flawed assumptions made that can result in a poor decision:

- Time savings = cost reduction
- Savings of labor is tied to products produced
- Overhead expense varies as a function of direct labor

Let us look at each one:

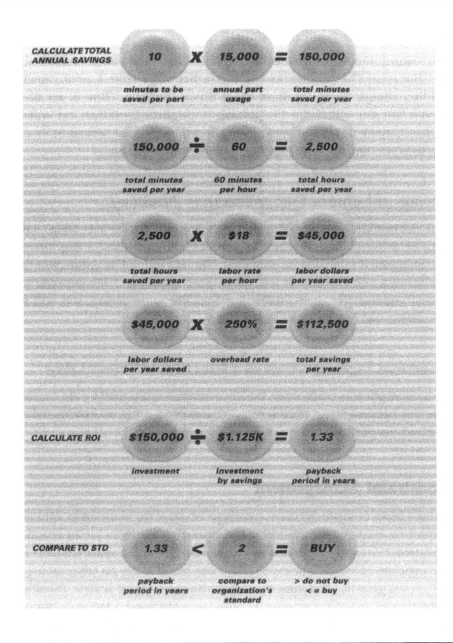

Figure 1.4 Investment Justification

Time Savings = Cost Reduction

This assumption is made when the per-part time reduction is multiplied by the labor rates. The unstated assumption is that when labor is saved, expenses will be reduced. There are times when this is not true. Time is *not* always money. You can reduce the time it takes to produce a product, but unless a worker actually leaves and payroll is reduced, there is no reduction in cost. The expense remains unchanged or is allocated elsewhere.

Savings of Labor Is Tied to Products Produced

This assumption is made when labor cost is multiplied by part usage. Labor capacity is not a linear function of production, but instead a step function of management's decision to add labor. The rise and fall of production does not dictate a one-for-one rise and fall in labor spending. If you examine your payroll expense and chart production over the same period, you will find correlation but not a direct relationship.

Overhead Expense Varies as a Function of Direct Labor

This assumption is made when the labor saved is multiplied by the overhead rate. Just because direct labor is reduced, overhead is not necessarily reduced in direct proportion.

Constraint Management

Using CM to evaluate decisions means to evaluate the decision in light of its impact on throughput, inventory, and operating expense. Let us revisit the investment decision illustrated in Figure 1.5. This is the same scenario with the opposite outcome. The expenditure was based solely on labor savings — no market growth. Therefore, the impact on throughput is zero. There is no reduction in operating expense, because we only saved 1250 man-hours — less than one man-year.[3] What is the impact on inventory? It increases $150K. With no additional sales and no reduction in expense, the decision is very simple. Find another way to invest $150,000.

Making decisions using throughput, inventory, and operating expense is not enough to fully evaluate a decision. How can I know that purchasing a

---

[3] Capacity is usually added in man-years, with one man-year equaling about 2000 hours.

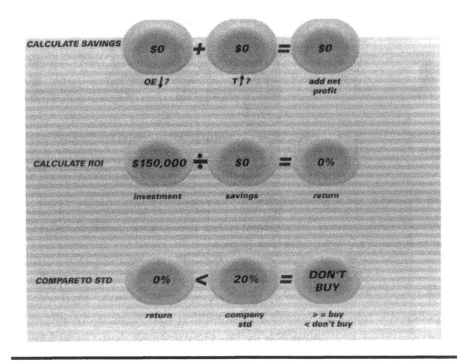

Figure 1.5   Investment Justification Using T, I, and OE

specific machine will make throughput increase? Who is to say that by reducing the amount of labor in one area, I cannot shift it to another, thus increasing throughput? In addition to measuring the system, we must arrive at a systemic understanding of how manufacturing works. If we do not understand how the system operates, we may change the wrong thing, resulting in a decrease in performance, or at best, no change in performance. We must understand the cause-and-effect relationships that govern the behavior of the manufacturing system. Constraint management provides that understanding.

Constraint management begins with one underlying assumption: **the performance of the system's constraint will determine the performance of the entire system.** This means that the worst performer will dictate the performance of your organization. Think of a chain. What determines its strength? The weakest link (see Figure 1.6). Manufacturing is analogous to a chain; each resource and function is linked. It takes only one element of the system to fail or to cause the entire system to fail. The weakest link is called the constraint. Goldratt defines the constraint as anything that limits or prevents higher system performance relative to the goal.

Figure 1.6   A Chain

The Nature of Constraints

There are three types of constraints: policy (paradigm), resource (physical), and material. Each has a different impact.

Policy Constraints

A policy is a rule, measurement, or condition that dictates organizational behavior. Policy constraints are, by far, the most prevalent (90%) kind of constraint and are the least expensive to fix. Many examples of policy constraints are found in *The Goal.* Batch sizing rules, resource utilization guidelines, and setup rules can all be considered policy constraints when they block the organization from achieving higher performance.

Policy constraints cannot be spotted directly, but a shortage of resources (material, machine time, etc.) will always point to one, which is why we usually think of constraints as resources and focus the implementation on resource utilization.

An example of a policy constraint was found at a consumer goods product manufacturer. This customer had over 200 finished goods SKUs. They had a machine that was costly to change over. In real dollars, it cost about $1200 in chemicals to clean. There were other expenses tied to setup time (about 2 hours). The machine was also very costly, being custom produced for this factory.

Because of these factors, the scheduling policy was to make long runs (at least a month's worth). As a result of the long runs, inventory of WIP and finished goods was very high. Simultaneously, customer service was poor, because many orders had to wait their turn in line — and the line was quite long. In order to alleviate the service problems, they introduced another policy called the "top twelve." If a product was in the top twelve sellers, it had the highest priority, pushing the low-volume products to the bottom of the list.

The distributors had very low confidence in the factory's ability to deliver anything on time. They padded delivery quantities and delivery dates. The factory had 3 months of work in backlog, with about 1 month past due.

The fix was simple. The constraint was the market, so they subordinated to its needs. The emphasis was no longer on optimizing the machine, but on delivering product. They changed the lot sizing rules to fill to a predetermined stock level and stop there. They changed the priority rules to focus on the delivery date of the customer. They asked the distributors to identify which orders were for customers and which were for stock. Within 90 days, past due was essentially zero, inventory dropped $3 million, and sales increased substantially. Costs rose, but not nearly as much as sales.

Resource Constraints

Resource constraints are much less common (8%) than policy constraints. In over 10 years of implementing constraint management, I have seen fewer than five real resource constraints. Most organizations have far more capacity than they realize. They do not see it because it is consumed and stored. The hallmark of excess capacity is excess work in process and finished goods inventory. If you have excess inventory, you have excess capacity.

Examples of resource constraints are machines, people, skills, and market. Mostly you will see policy constraints show up as resource constraints. Usually the reason for a shortage of resources has to do with the policies governing the utilization and acquisition of those resources. *The Goal* is a good example of this. You never saw any fundamental change in the manufacturing plant discussed in the book except in the management of it!

Material Constraints

Material constraints are the least common. These are usually things such as scarce materials, but can also be a commonly available material that is in short supply due to problems in the supply chain. Several years ago, after a fire at a major computer chip manufacturing plant in Japan, chips were in short supply. Computer manufacturers reacted by offering high-end machines that were loaded with profitable options. This was their way of exploiting the shortage of a critical component. They just stopped making the basic, low-profit machines.

The distinction of whether a given constraint is a policy, resource, or material is mostly academic. Your concern as a manager is how best to manage the resources available to achieve the goal. I have included distinctions to point out that most of your constraints are not resources, but are policies that govern resource utilization. Therefore, the implementation is focused on policy and procedure development, rather that process improvement.

Focusing Steps of Constraint Management

Because the organization's performance is limited by its constraint(s), our process to manage the chain must ensure the constraint is the focal point of everything we do. The following steps allow us to focus:

1. IDENTIFY the system constraint(s).
2. Decide how to EXPLOIT the identified constraint(s).
3. SUBORDINATE everything else in the system to step #2.
4. ELEVATE the system's constraint.
5. GO BACK TO STEP #1.

Identify the System's Constraint(s)

Identifying the system's constraint(s) is essential. If you do not know the location of your constraint, you do not and cannot understand your business. The constraint determines many aspects of your business: output, profitability, return on investment. The identification of the constraint is the beginning of your improvement efforts. If you understand where the constraint is, you will leverage your improvement efforts. You will make better capacity expansion decisions. You can begin to use your manufacturing asset more strategically in the marketplace. If you do not understand your constraint, you are often guessing.

In a traditionally run manufacturing environment (where there is a great deal of excess capacity), this step is relatively easy. Where is the largest queue of work in process? The answer to this question may produce the location of the constraint. Of course, in a traditionally run manufacturing business, there are piles of work everywhere. There may not be a single location with the largest pile of inventory. So, the goal here is to "convict" a resource of being a constraint, to build the case for a resource being a constraint. The idea is that no single effect will conclusively lead you to the constraint (although it might). What we want to do is to build a case and a logical argument to decide where to start the implementation.

Do not worry too much about correctly identifying the constraint at first. As you will see, even if you are wrong, the performance will still be limited by the constraint, and by focusing in the "wrong" area, the "right" area will show up immediately. If you can live with a little ambiguity, the approach is workable. I ask these questions:

- Where is the work backed up? Where are the piles of work waiting?
- Where do most problems seem to originate? In an assembly environment fed by a machine shop or multiple plants, the constraint will show itself often on the assembly shortage list. So, ask the expediters where they go all the time for parts.
- Are there any resources with high utilization? In many plants, a lot of effort is spent tracking resource utilization. Which one(s) have the highest? If it is a constraint, it never runs out of work and is always behind. Someone has that figured out and keeps it running.
- Is there a "key" resource? This one can be misleading[4] and should always be taken in context with the other effects. Sometimes, the production leadership has decided that certain resources are "critical" or key resources, and if they do not keep that resource operating, the other resources will starve.

The effects of the constraint are always blockage of the resources preceding the constraint resource (resulting in a queue) and starvation of the resources downstream of the constraint resource (Figure 1.7). If your suspected resource doesn't exhibit these effects, it is not the constraint.

Some other questions you should ask are:

- Can other resources, when staffed at plan, outproduce the suspected constraint? If not, you have a new suspect.
- If we add another resource, would output of the facility increase?
- When this resource is starved or idle is the entire production plan thrown off?

The answers to these questions, when take as a whole, will give you assurance that you have indeed found the constraint. Remember, even if you are wrong, you will soon identify the real location and nothing will be lost. I have found that when the location of the constraint is not clear, and we pick the wrong

---

[4] I say this effect can be misleading because I worked with a production manager who was convinced a particular piece of equipment was a constraint, yet it only affected about 20% of the product and experienced frequent downtime with no loss of revenue. He did not understand his process very well.

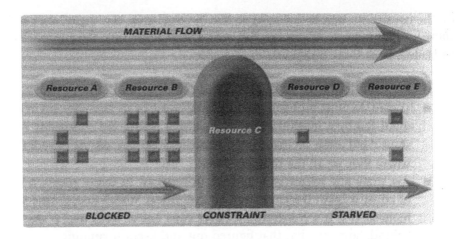

Figure 1.7 Starvation and Blockage

location, the correct location shows itself very quickly. I am then very certain of the location of constraint.

The Value of the Constraint

The throughput of the entire system is limited by the constraint. Therefore, we can measure the impact of a single decision based on what happens at the constraint. As with a funnel, the rate of flow through the funnel is determined and measured at its narrowed point. The constraint acts in a similar way to the organization (Figure 1.8). This substantially simplifies decision making, because your entire decision process can be boiled down to one question; "Will this make the funnel larger?" (Will this make throughput increase?).

Look at the financial impact the constraint has on your business. Figure 1.9 shows how dramatically an hour lost at the constraint resource can affect the bottom line of the organization. In this organization, every hour lost at the constraint resource reduces profit $14,706. Usually, we think that each resource acts independently on the bottom line of the organization. That is why we end up with chargeable rates of $83 per hour or some other nonsense. The constraint governs profits. Nonconstraints have so little impact we do not bother to measure it.

Figure 1.10 shows how the 80/20 rule is skewed when we are dealing with constraints. There are many things you can do to improve profitability. Each thing will have some impact. The 80/20 rule says that 20% of these things

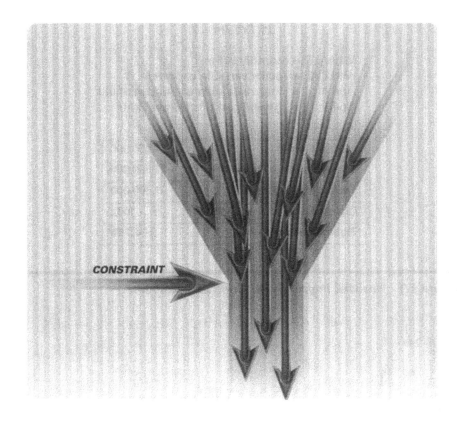

Figure 1.8 Funnel

will yield 80% of the results, but constraint management says that there are few things you can do to have a significant impact; the rest have a minimal impact. The 80/20 percentage changes to 99/1.

To see if you are working on the constraint, look at the history of net profit and return on investment. If profitability is constantly rising, the constraint is always being elevated. If your profits are not growing, you are not paying attention to the constraint. You can ignore it, but it will not ignore you.

Decide How to Exploit the Constraint(s)

Once you have found or decided where the constraint is or should be, plan to make the most of it. All sources of wasted time at the constraint should be eliminated (assuming you have a resource constraint). As shown in Figure 1.9, an hour lost at the constraint has a huge impact on profitability.

$$(S-TVE)/H = T$$

Where **S** = Monthly sales
And **TVE** = Totally variable expense
And **H** = Hours available at the constraint
And **T** = Throughput (Profit) per hour

Monthly Sales	$4,000,000
Totally variable expense	$2,500,000
Throughput value add	$1,500,000
Constraint hours available	1032
Throughput per hour	**$14,706**

Figure 1.9   Financial Impact of the Constraint Resource

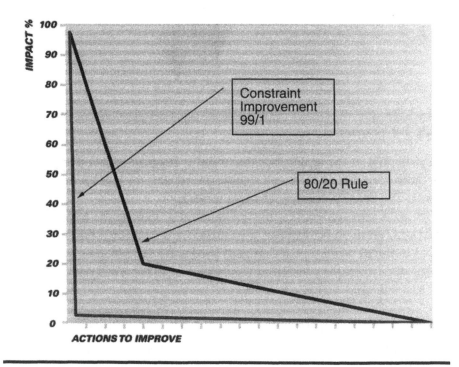

Figure 1.10   Relative Improvement Using CM Focus vs. 80/20 Rule

The most obvious source of wasted time is downtime. It is the easiest to find and fix. Eliminate all idle time at the constraint resource. This, in itself, will have a dramatic impact on the bottom line. In assembly plants, this may be as simple as eliminating shortages at final assembly (well, simple is a relative term).

Less obvious is the time wasted in production. Are you producing inventory to be sold in the next week? Are you being "efficient" with setups, producing inventory for future sales, while current orders are forced to wait? Do you have the best operator running this resource? If, indeed, this resource is the constraint, it makes sense to have the very best person running it.

I had a client that produces screen-printed graphics. The constraint was the press line. The problem was they were constantly down and didn't have enough press operators (six operators on the line, each manning a different press). Throughput was being lost and orders were late.

When we investigated the compensation structure, we found that this position, though highly skilled, was the lowest paid in the shop. People would leave at the first opportunity to get a better paying job in the plant or elsewhere. When we changed the pay structure to make this position the highest paid in the shop, the turnover dropped dramatically and throughput went up.

Remember that the constraint resource limits the performance of the entire system. Productivity there increases productivity for the entire system. A small investment at the constraint will have a huge payoff on the bottom line.

Subordinate Everything Else to the Constraint(s)

Once the plan is made for the constraint resource, ensure the other resources are working toward the same plan. This is synchronization. Your decisions to maximize the system's constraint are your decisions to maximize the entire system. If you do not synchronize the nonconstraints to the constraints, then you will find yourself very frustrated keeping the constraint busy with the right product.

Subordination can take many forms, especially with policy. The most common form of subordination is ensuring that material release matches material consumption at the constraint resource. Release just what the constraint needs — no more, no less — and you will have de facto subordination of the nonconstraints.

Elevate the System's Constraint

When you are sure the weakest link has been exploited fully, look for more capacity (equipment, people, time). Additional investment should be considered only after you are certain you have done the best you can by breaking the policy constraints.

Go Back to Step 1

Eventually, you will break the constraint and a new one will appear. If you want continuous improvement, you must continuously identify and break constraints. Remember that the rules of yesterday became the constraints of today.

Do not allow inertia to become the constraint.

2 Implementation Overview

The Approach

Before getting into the detail of the policies and practices, I want to give you the general approach that shapes the decisions and implementation priorities.

- Getting control of the business is paramount. Promise and deliver orders on time.
- Get the implementation going as quickly as possible with concrete changes that demonstrate results. Do not wait for perfect solutions before you act.
- Make changes as soon as problem policies are identified. Break policy constraints as they are identified.
- Start with simple, low-tech solutions first, then implement refinements later. Automation is necessary, but useless without understanding what automation does.

I always strive to prevent localized implementations with a single person who understands the overall system. What I strive for is nothing less than a complete change in the behavior of management. Therefore, we must have a critical mass of managers who understand, embrace, and practice the philosophy.

Unlike many cultural changes, *total* buy-in from all managers is unnecessary, only a general agreement not to hinder the project. As long as people are willing to try, I know the changes will work. Once the system is operational, they will not want to go back to what they were doing before.

Get Control of the Business

Mastering the order fulfillment process is foundational for other improvements. It means that you consistently deliver product on or before your promised delivery date at least 90% of the time. If you do not know what your company's on-time performance is, you should assume it is less than 50%. And just because your customers are not complaining it does not mean that your performance is good enough. Not having mastery of the order fulfillment process in manufacturing is like a restaurant that cannot deliver the meal. Who would patronize a restaurant that did not consistently deliver the meal in a timely fashion?

Understand that on-time performance is a competitive edge issue. The competency can be easily copied. If the market is dominated by a poorly performing organization and I want to take the market, all I have to do is develop a more responsive order fulfillment process. I do not need a higher quality product, a more advanced product, or even a cheaper product. If I care for your customers better than you do, they will be mine. You do not have to look far to validate this truth. Examine your own behavior. Do you not prefer to take your business to the merchants that serve you better — even if the price is higher? Your personal aggravation factor always figures into the buying decision. Despite the purchasing agent's protestations, emotion still figures into the buying decision. Examine the psychology of the buyer. Their first priority is not to shut production down — get the product there on time. If you fail to do that, you have just created a major headache for him (or her). She would like to keep her job and avoid problems, too. Price becomes an issue only when two competitors have comparable service.

Delivery reliability is the single most valued dimension of customer service. Improving this single dimension will increase sales and increase the margin of existing sales (Figure 2.1).

The way to get control of the business is to synchronize the activity of the business to the customer. By synchronizing the business, lead time is reduced, which affects margins and sales and, of course, the bottom line. That is to say nothing of reducing the waste associated with poor resource synchronization — expediting, premium freight, and overtime. This should be done before major marketing initiatives. If you cannot effectively manage the level of business you have today, imagine how difficult it will be with 10, 20, or 30% more business. Poor synchronization is often the source of the problem where organizations have increasing sales and profits, but not at the same rate (Figure 2.2).

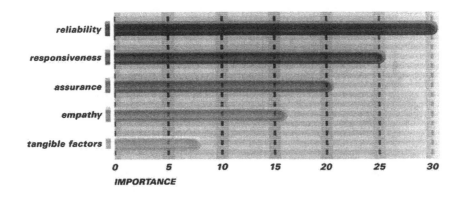

Figure 2.1 Service Quality Factors (Source: Forum Corporation, Boston, Customer Focus Research Report)

Figure 2.3 is the current reality tree of most order fulfillment processes. The way to read it is with each arrow leading from the cause to the effect. Read it like this: IF "Resource activation is good," THEN "People will work on any job available." IF "People will work on any job available," THEN "People work on the wrong orders." Take your time and read the entire tree.

The bubble at the top, "The prevailing belief is that 'Resource activation is good,'" means that out of this belief, the entire order fulfillment process has been designed. It is this belief that must be destroyed in order to synchronize the business successfully. "Resource activation is good" is another way of saying that unless the resources are working, we can't make money. Conversely stated, how can we make money while many people sit

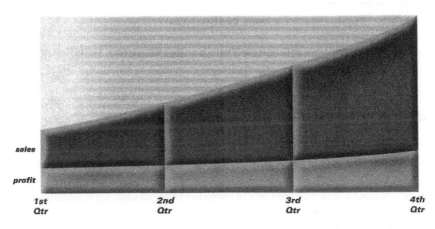

Figure 2.2 Sales and Profit Rates

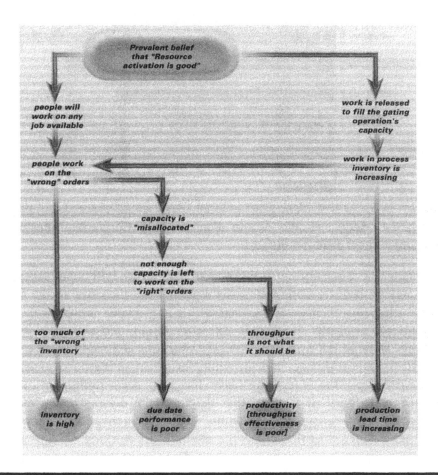

Figure 2.3   Production's Current Reality Tree

idle? Activation equals efficiency. This is a deeply rooted belief — cherished by all.[5]

Get the Implementation Going Quickly

The largest obstacle to implementation is organizational inertia. If people are not actively engaged in improving the business, we must engage them.

---

[5]   I do not think I have ever been able to completely kill this idea. My approach has been to recognize that I cannot, so I use a sort of Judo — use this idea, but channel it in a different direction. Some people think that managing according to the constraints means that people and/or resources will be idle. Later, you will see how to channel this excess capacity and take advantage of the fact that people MUST be busy. Rather than move work to people, we will move people to work.

There is quite a bit of work to be done to implement CM and, at times, it will be difficult. Therefore, we have to achieve momentum early in the project. We do not want to overanalyze before we act. Some very simple things can and should be done immediately to start the ball rolling.

When you reexamine the business processes, you will undoubtedly find stupid things. This is true even if you are not implementing CM. Systems and policies are implemented for good reasons. Time passes and the situation changes. The original reason for the policy is no longer valid, but it stays in place. The effect of the policy is that it is no longer helping things, but blocking increased performance.

Start Simple

I could also say "start small." Begin with simple systems and procedures that will yield *some* results, rather than trying to implement a sophisticated system or procedure that will yield optimal results. Try using a ratio of time to implement vs. time to use. If you cannot start doing it in a week or so, do something else that can be done in a week or so. Save the hard stuff for later in the implementation.

You are trying to break the inertia, change the culture, and teach a new way of doing business all at the same time. So aim for changes that can be put into practice quickly. This allows people to see results and learn the why of the philosophy. It is also low risk. You do not need to spend weeks doing a study to see if it works — try it and make adjustments. Change is happening, and it is happening in very small increments, so the resistance to change will be very low.

One-Person Implementations

I have often seen an organization bow to the will of one person (when he is the general manager). A single individual can implement a partial solution and get substantial results. The problem occurs when that person leaves. I watched one gentleman move from company to company, implementing CM at each, creating breakthrough results. He then left, and the implementation left with him. His replacement, not being from the original team, would dismantle the implementation with predictable results.

The objective is to anchor the philosophy and practice deeply in the organization. It *should* be difficult to change. However, in order to do that, a greater investment in developing the management team is required. That takes time. It means that the results will be slower in coming, but will be longer in duration. Take your time and develop the team. Do not try to do the implementation by yourself.

Consensus Building

We do not need everyone to be a CM evangelist to be successful. Not everyone has to be fully convinced that the process will work. What we do have to do is create *some* champions and prevent *active* opposition from the rest of the organization. If we can get people to try, the results will speak for themselves. Remember that the approach is based on small, incremental changes with measurable results. At first, we are not asking for the world, only a small piece of Rhode Island. As momentum increases, so does buy-in. Do not worry if people do not jump up and down ready to change the world when you first suggest it. You know that this is not the first trend to come along.

Elements of Success

There are two main contributors to success — good system design and the implementation process. These are the two most fundamental problems with any initiative: "what to change to" and "how to orchestrate the change." You will find many good ideas, but you must be successful in implementing them. A poorly conceived idea implemented well will not work, nor will a well-conceived idea that is poorly implemented. The things to make the implementation fail are almost innumerable, but the reasons for success are few.

Solid System Development

A solid system is robust. It is simple and it is practical. It may or may not include the use of the computer system. My experience dates back prior to the proliferation of APS (advanced planning and scheduling) software products, so by necessity, I had to make do with manual systems and manipulation of the MRP/ERP system to do what needed to be done. Many people will insist that before you apply software, you should focus on the business processes and the behavior of management. I do not feel so strongly as I used to, because I have seen environments where software and process were applied simultaneously and were successful. But I concede that a poorly conceived information system will not help and will create larger problems than you had when you began. A conservative approach is to focus on the process, then automate when the limits of the manual system are reached.

Planning

The planning process is built around recognizing demand and reconciling it to the available capacity. It covers the medium and short term. Typical decisions are related to production and evaluation of vendor capabilities (make vs. buy).

In long-term planning, synchronization is not the issue; reconciling market opportunity to capital availability to break constraints is the primary concern. Typical decisions are related to capital costs such as purchasing machines or new market initiatives.

Execution

The execution process concerns itself primarily with satisfying the plan and dealing with variation. Out of the execution process will come valuable information that will be used in planning. The execution processes help you understand and manage the internal resources.

Organizational Alignment

Organizational alignment is when all elements of a company work together in concert within the context of the organization's core ideology and type of progress it aims to achieve — its vision or goal. Let us break this down a little bit. Alignment is achieved when all elements of the company:

1. Work together in concert
2. In the context of its core ideology (culture)
3. To achieve its vision

The new system cannot work at cross-purposes to existing systems, culture, and vision. The constraint management system is almost solely focused on increasing throughput with the reduction of cost and operating expense the secondary consideration. Therefore, if your organization's culture gives primary consideration to costs, you will find that implementing constraint management fully in the organization will be extremely difficult, because the cost-saving mentality is focused on the independent links of the organization. Constraint management focuses on the overall system, and is not concerned with the suboptimization of the individual links within the chain. The goal of constraint management is to optimize the performance of the *entire* system.

The implementation strives for organizational synchronization. So creating processes that promote organizational alignment is at first geared toward organizational integration. Later phases of the implementation will affect greater parts of the organization, and synchronization will be achieved not by scheduling, but through behavior modification. You will see that your organization will positively change in ways you cannot fully appreciate — not until you are deeply into the implementation.

Organizational alignment addresses these problems.

- Behavior inconsistent with your core values and purpose
- Organizational structure that inhibits progress
- Goals and strategies that move the company away from its basic purpose
- Policies that inhibit change and improvement

You thought you were just going to implement constraint management? I am trying to point out that the constraint management system will have far-reaching, positive consequences. Many of these consequences are side effects with no extra effort required. I am getting excited about your implementation just writing about it!

Implementation Process

The process of implementation has to do with how you manage change. There might not be much new here, all I can do is explain how I do it. Rest assured the process is repeatable. I have used and refined it many times. Many of my former customers and employers received the "benefit" of my experience. I made many mistakes. I am not promising a perfect process, but it does work. But it is critical that you pay close attention to how you manage change. Done correctly, you can overcome resistance and work toward success. Done incorrectly, the implementation is doomed.

The CM Production System

A system can be broadly defined as a collection of policies, procedures, and behaviors working together to achieve a specific outcome. The CM production system is concerned with maximizing the output and value-add of the constraint resource(s) and synchronizing the business to that resource. It is achieved using a scheduling system called drum–buffer–rope (DBR). DBR scheduling is focused on three areas — the constraint (drum), gating resources/material release (buffer), and shipping (rope). There may be other scheduled areas. This will depend on how many independent throughput channels exist and how much protective capacity exists in the system.

There is an emphasis on establishing and maintaining buffers at strategic locations in the company in order to manage variability. Despite the presence of buffers, work in process and finished goods inventories are very low, compared with other techniques. The workforce is flexible, with cross-functional expertise common. People move freely among departments and resources to resolve peak load conditions.

The system is broken into four subsystems:

1. Production planning and execution
2. Material planning and execution
3. Business planning/master scheduling
4. Organizational alignment

Production Planning and Execution

The production planning and execution system concerns itself with reconciling current demand and current capacity while maximizing throughput effectiveness. Throughput effectiveness is the ratio of throughput to operating expense (T/OE). This recognizes the impact that production has on the throughput of the business and the trade-off that exists between maximizing throughput (revenue) production while minimizing expense (cost). In light of the focusing steps, this system is concerned with maximizing the constraint and subordinating the production resources to that constraint. Generally, this system "drives" the others, translating the strategic objectives to the tactical actions.

Production Planning

The heart of the system is a daily production plan — the drum (or drum schedule). The daily schedule is the plan for maximizing the throughput of your business and is the focus of your scheduling effort. Remember that the performance of the entire system is limited by what happens at this resource. If you have no schedule at all and your lead times are very long, you *could* schedule in weekly buckets to start, although it will not be long before you see the need to schedule in daily buckets. All other activity is subordinate to this schedule, the drum.

The ERP/MRP system can be manipulated to give a rough schedule of the constraint. You will find that the scheduling logic in most systems can be used to roughly synchronize your plant. It is not optimal, but there will be a dramatic improvement in your ability to plan and execute.

Accurate Demand Information

It doesn't help much to have a wonderful schedule, if the demand being scheduled does not reflect the needs of your customer. Start scheduling on a lot-for-lot basis. Strip all EOQ-related and lot-sizing rules from the MRP system except what the customer needs (lot-for-lot). Usually, when the project is beginning, there is one of two problems. Lead time is too long or delivery performance is poor and there are lots of late orders. Stripping the lot sizing rules will be effective in reducing lead times and improving on-time delivery performance.

Make sure that you are really making customer requirements. In a stock replenishment or distribution environment, it is often difficult to distinguish between the needs of the customer and the replenishment order from the warehouse. The planner/scheduler must have the ability to distinguish between the two if he/she is to plan the constraint to maximize throughput and customer service. Remember that it is not throughput until it is sold. Therefore, maximizing throughput means real orders get higher priority when needed.

If you must replenish inventory, make sure that the priorities between replenishment and customer orders are established. It does not make sense to make stock when you have customers waiting for product. They want to give you money! Don't make them wait!

To develop an accurate schedule, you must have a good picture of your work-in-process status. You do not need to know where everything is, just its location relative to the constraint. Has it been processed or not? In this way, you can identify the net capacity requirements at the constraint.

Make sure your open order database (sales orders and job orders) is accurate and free of orders that are not needed. If your database is already clean, great. But most organizations use an informal approach to planning and have not developed the disciplines to manage their data. This is essential to identify the true item demand, which translates to the true capacity demand on the constraint resource(s).

Order Commitment Process

The order commitment process is the way you determine when you have fully identified the customer needs and are ready to produce the product. If you are certain you do this, great. Skip to the next section. I include it because I find this issue so often in make-to-order environments. The problem looks like this: An order is launched to production without knowing all material and delivery information. As the order is produced, questions arise about configuration and delivery. As these questions are answered, priorities are changed due to material availability, additional processing, or customer requirements. The effects of this problem are partially completed jobs on the shop floor and late deliveries.

The implication of this problem is that valuable time on the constraint is wasted. Either time is put into a job that must wait, or the order is reconfigured and must be redone. Both have a negative impact on throughput and operating expense.

The mind-set we must encourage is that once an order is started in manufacturing, it must be completed. It should not stop for any reason. It is better to have finished goods and have the product off the shop floor than to have work in process cluttering the plant. When the project is completed, your lead times will be at least half what they are now and the velocity will be much greater, so ensure the product is made correctly the first time.

Product/Order Configuration

The product must be fully designed before you release it to manufacturing. You do not want your people in the plant to design the product. Do not interpret this as a slam against the people in production. If you want them to engineer your product, great! The issue is being able to control the variation in your process. If you do not specify the product, you will get something different each time, wrecking the schedule. You will have a hard time predicting delivery dates and costs, with predictable effects on your ability to sell and make a profit.

Your definition of *fully designed* may be different from mine. I can foresee some instances where not all product specifications can be completed before production begins. If this describes your organization, you will have to build additional buffer time into the production lead time to account for the increase in uncertainty. Not an ideal situation, but you *can* compensate.

Resource Availability

In addition to designing the product before it reaches the shop, you have to validate the resource availability to produce it. In terms of capacity, this means quantifying the order's impact on the constraint. In terms of material, you may not need to look at every part, but certainly should consider those critical materials that are difficult to get in a short time.

Unless you can quantify the impact on the constraint, you cannot plan. If you do not know the hours, try using revenue or pieces. Most companies have excess capacity and so, initially, planning can be done roughly with positive results. The main idea is that resource consumption for each order must be identified in order to plan production. We do not need to quantify everything, and we do not need to be precise to get results. We do need to identify the impact on the constraint of accepting the order.

Standard Lead-Time Policy

Order delivery commitments must be made based on actual backlog at the constraint. Because there is a lag between the market and production planning, the sales organization needs a tool to promise customers delivery dates. The standard lead-time report is the way manufacturing communicates production load and status to the sales organization. This report should be updated weekly at a minimum, more frequently if sales or capacity is changing more frequently.

This report is an excellent way to inform management when capacity is out of balance with market needs. If lead times become an obstacle to sales, this then leads to a conversation about capacity and how much do you need, which is infinitely preferable to a conversation about why production is always late.

Schedule Execution

A schedule is only as good as your ability to execute it. The most problematic component of the CM system is execution. It is at this time that behavioral change becomes an obstacle. We are going to ask our people to focus on the constraint, not their own area. They will have to suboptimize to meet the schedule at the constraint. The definition of what is acceptable performance is going to change, and it will change dramatically.

The behavior we want to create is to maximize the constraint and subordinate everyone else to the needs of the constraint. Therefore, acceptable performance at nonconstraints is no longer maximizing individual productivity. Instead, acceptable performance is judged on how well you supply product to the constraint in a timely manner. Of course, the constraint will still be judged on the traditional measures of efficiency and production. Do not underestimate the difficulty in changing this mind-set.

Daily Buffer Status Meetings

The buffer status meeting is when you evaluate the content of the buffers, identify problems, assign responsibility to fix the problems, and measure performance of nonconstraints. It is a way of resynchronizing the functions of the business to the plan.

The meeting as I have designed it is a 15-minute information exchange. It is NOT a problem-solving session — problems are solved off-line, where only the people required to solve the problem are present. The plan vs. actual is the only topic of discussion during this information exchange. Accountably for resolution is assigned then we move on to the next issue. All buffers are reviewed — shipping, resource, and assembly.

Specify Queue Areas

The plant should run on a visual basis as much as possible. It should be obvious when there is a problem by watching the queue areas. When an area falls behind, the amount of work in the queue rises, providing a visual indicator for action needed. When the queues are empty, a different kind of action is required. Perhaps we move the people or look upstream in the process to see what the problem is.

The queue areas provide a visual cue of problem areas that communicates to anyone on the shop floor. If people are trained properly, they will take independent action to resolve the problem before it affects throughput. The queue areas are a measurement — prompting behavior that promotes synchronization with the constraint.

Material Planning and Execution

There has been quite a bit of discussion about material constraints, especially among the Advanced Planning and Scheduling software firms. The purists say there should be no material constraints. In principle, I agree. You do not want the throughput of your company to be determined by your vendors, over whom you have very limited control. However, they do support the constraint and should be subordinate to your drum schedule.

ERP and MRP systems are designed to accomplish just this task. Your job will be to tie the execution plan to the material requirements plan. If you do not have or use MRP, then you will be establishing some sort of reorder point and replenishment system to ensure adequate stocks of raw materials. Whatever method you choose, you must ensure your vendor is not the constraint. Never allow material availability to affect plant throughput.

Business Planning/Master Scheduling

A master production plan based on the constraint will help you avoid shortfalls in capacity. The master schedule is merely a rough-cut capacity plan of the constraint resource for the next 3 to 6 months. It is important that the schedule reflects the actual constraint output, but then that is true of all schedules. The planned capacity of the constraint should be based on demonstrated capacity. It must be a realistic plan, conservative in its estimates of available capacity. This will help avoid the problem of overcommitting the plant.

The purpose of the master plan is early identification of problems. Because you are planning work that you probably do not have yet, a rough-cut plan is acceptable. The only difference between the traditional master schedule and one that is constraint based is that resource availability is gauged by constraint resource (drum) capacity rather than by overall planned hours.

I have seen the market overwhelm the capacity of a plant implementing DBR, and then DBR gets the blame for the lack of responsiveness. Your market *will* respond to better performance by giving you more work. Implementing this process early will give you a mechanism with which to respond before the plant becomes the constraint, thus taking advantage of the wonderful sales opportunities that will be presented to you.

Organizational Alignment

The new production system must not conflict with the rest of the organization. Part of the implementation effort will be integration; part will be changing the larger organization. This will be done mostly through the measurement system you will employ, but you will also have to educate other parts of the business that interact with the new system. Others need to know what buttons to push and what levers to pull to accomplish the needs of the customer and shareholders.

Measurements

The measurement system is designed to create the desired behavior, diagnose problems, and track progress. You will want to measure the output of the constraint resource on a daily basis. You will want to know how effectively the nonconstraints are subordinating to the constraint. You will want to keep people focused on the constraint and its impact on the global objectives of the business, so you will produce a daily shipping report and a daily booking report. In some cases, you will stop measuring some things because you want to deemphasize those items.

Education

A basic form of alignment is education. The organization will be blind to the existence of constraints and how to respond to them. Not everyone needs to be an expert, but if we are to guide decisions based on constraints, we will have to teach those in the organization what a constraint is. We have to give the organization at large a basic understanding of constraint management principles.

When we introduce the new measures, we will have to teach people how to effect them. I am not interested in frustrating people with metrics they cannot change. Many of these will be new to people. So, we have to teach two things — the meaning and how to respond.

Education also performs a basic behavior modification function. As new information arrives, people will change their behaviors. However, the behavior change comes only when the measures reinforce it. So, we need both — measurements *and* information — to change the behavior of the organization.

The Implementation Process

Now that you see the components of the implementation, this is how they work together. I will explain it sequentially — the prerequisites, if you will. The implementation process is broken into five components:

1. Consensus building
2. Assessment
3. Design
4. Implementation planning
5. Project management

Consensus Building

The implementations that have the best, most enduring results are led from top management (the person(s) most accountable for profit and loss). The most practical reason for this is that top management has maximum control over the resources and the implementation *will* consume resources. The second reason is that the organization looks to senior management for direction and guidance. If they are not on board, it will be much more difficult to get people to risk changing the process. The third reason is that they control the measurement process. If you cannot control what gets measured, you will be unable to change the behavior of the organization.

Do not confuse management leadership with management implementation. We do not necessarily need to have the CEO or president directly involved in the implementation. My experience has been that he or she does not know or want to know the details enough to be effective in leading the project. Besides, they should not be focusing their attention on this level of detail; they should be building the business and the market.

Once the top manager is supportive, then I typically progress to the next level of management. The kind of implementation process I have described cannot be done without the support of the senior management staff. Depending on the culture of your organization and leadership style of the top manager, you may not be able to get support from the president or general manager unless you have broad support on the staff.

From there, you progress to the middle and front-line leadership and finally to the direct labor people. The process of building consensus I have described is top down, because that is the way it works best. So — how to get the top guy? What if you do not have him on board? You can implement some things that will earn you the right to be heard, but unless he is on board, you won't achieve a lasting implementation.

How to Do It

Consensus proceeds from interest to inquiry to commitment. If your organization knows nothing about constraint management, start with *The Goal*. Start passing out books to decision makers and opinion leaders in the organization. Send out articles of success stories about constraint management. Pick the ones that concern the same problems as your company has. Talk it up. Attend constraint management seminars and workshops to educate yourself. Organize training sessions for your people. You can do this yourself or hire a consultant.

Once you have sufficient interest, organize an internal workshop for the senior management team to learn what constraint management is about. There is no substitute for this step. Do not forget the magnitude of change you are asking people to make. They should invest at least a day to learn in detail what it is about. *The Goal* is too simplistic and narrow to get people to commit. Someone who is knowledgeable about constraint management concepts and experienced in implementation should lead the session. Management will have some tough questions; be prepared! After the workshop, conduct a study mission to a business that has implemented constraint management successfully. There are plenty of success stories and most people are happy to share their success.

This step, if done properly, should lead to commitment. However, commitment will not come unless you have a proposal to achieve a specific result. The proposal should be tied to quantifiable, bottom-line results. The initial commitment may only be a pilot project. If that is what it is, take it and run. The results will be obvious to anyone and will lead to greater levels of commitment.

To summarize, get people's interest using *The Goal*, articles, and discussions. Create opportunities for further inquiry using workshops and plant visits. Secure commitment to a specific project.

Champion

An overlooked part of building consensus is maintaining momentum. The job of keeping everything together falls to the project champion. He or she is the one responsible for maintaining forward movement when things go slowly. The champion is the person who takes personal responsibility for the implementation. In essence, the champion plays, "you bet your job" on the outcome of the project. Without someone's taking responsibility for success, the implementation is doomed.

Set Objectives for the System

I cannot emphasize enough the importance of having concrete objectives established for the project. The project must have a direct impact on either your customer or financial performance. It must reduce cycle time or improve on-time delivery, improve manufacturing productivity, or increase throughput. The objective must be tied to the financial or customer performance if you are going to achieve lasting, broad-based support.

Measurements of the Objectives

The objective must always be measurable. In order to measure the project performance, you must clearly articulate the impact the project is going to have on the business. If you cannot do that, why do it at all?

The measurement is also a gauge of implementation success. It sets you up for a "win." It helps maintain support for the project. I have seen projects stall, even though they were doing what they were designed to do, because the organization did not have an idea of how well it was doing. If you are not measuring the performance, the project can be killed before it is implemented.

Assessment and Design (Identification of the Constraint)

Identify Areas of Focus

The initial stage of the implementation is walking through the five focusing steps (identify, exploit, subordinate, elevate, and go back to step 1) and painting the picture of the business after the project is completed. In reality, the implementation is never finished. The implementation plan will walk the organization through the focusing steps once or twice, then the project manager should step back to assist only where necessary.

To identify the constraint, first chart the order fulfillment process. During this process of interviews and analysis, look at what is being done at each step. But this is only a means to an end. The real goal is to find out *why* things are done. Since 90% of constraints are policies, you have to know the *why* before you can address the *how*. Therefore, the implementation will include a detailed analysis of the current processes.

Implementation Vision

Once you have the constraint identified, skip two steps and go directly to elevate. I contend that every implementation begins with policy constraints. Why exploit and subordinate an erroneous policy? You should break it immediately.

Breaking the policy constraint is easier said than done. It is not enough to say, "Stop that!" You have to offer a compelling alternative. Spend time developing a visual overview of how the business will work when the project is completed. This gives you and the project direction.

Your approach to designing the system can be done in two ways. One is to identify, then develop a system to exploit the predicted capacity constraint. The other approach is to decide where the constraint *should* be, then break all other potential constraints that prevent you from keeping the constraint in the desired location.

Some people refer to this process as identifying the "control point," rather than identifying the real constraint. After a time, the "true" constraint will be best managed from a strategic standpoint. You may as well begin that process early.

From now on, I will use the terms *constraint, drum,* and the *control point* synonymously. You will find that *drum* and *control point* will become the common terms for the internal constraint (while you have one), and the "real" constraint will be called by its true name. Most times, the constraint ends up in the market, right where we want it.

Implementation Planning

Before you move forward, you should have a sense of how the implementation is going to be accomplished. Develop a plan of *how* you are going to implement. Topics for consideration include organizational culture, political power bases to tap (or negate), the position of the potential project manager and champions, organizational sophistication, and availability of information technology services. The successful project plan will allow for change management as well as technology management. The technology is not very difficult; it is managing the changes that will prove the most challenging.

Preparing the Organization for Change

At this point, we have the top managers and key staff educated. We have evaluated the organization, developed a solution and a plan of attack. Next, we have to mobilize the organization and present the analysis, solution, and implementation plan to management before moving ahead. They will have already bought into the idea of changing the direction of the company, now they have to buy into the specifics of what is going to be changed. This is a very sensitive presentation. Somebody's sacred cow is going to be killed (probably more than one). People do not like being told their baby is ugly (sorry for the mixed metaphor), even when they ask your opinion. Present your proposal and ask for commitment. Afterwards, have someone you love pull the arrows out of your chest.

Compared with the presentation to the management team, gaining buy-in from the rest of the organization will be easy. They already know things are messed up and are willing to change if they believe it will be better when they are done. Your implementation will not affect the actual work practices of most of the people, but the environment will be changing, and people will need to be alerted to avoid a backlash.

The project launch consists of formation of the project team, a letter from the president to the organization announcing the project, and a brief meeting with all employees to announce the start and answer questions. This is where the real commitment lies. The president or general manager tells the organization this is going to happen and who is responsible for making it happen. It is very difficult to go back after this step. Plus, people get cool T-shirts.

To summarize, this step is a way to continue to build commitment toward changing the system. It does not make you any money, but it will eliminate resistance when you begin to actually change things. This is what you have to avoid.

Project Management

The project should be structured in such a way as to involve as many people as possible, but no one should be on the project team unless he or she has a specific job to do. Everyone on the team should have a specific task or responsibility. That includes the president or GM. Communication takes place via the measurement reports, procedure approvals, and the steering committee meetings. Part of the implementation plan is to prepare for the project planning, tracking, and reporting aspects of the implementation. While not affecting the procedures or processes being implemented, effective project management will have a dramatic impact on the success of the implementation.

A "Typical" Implementation

Your implementation will generally follow this pattern: procedure development, education (conceptual and technical), implementation, OH MY GOD!, procedure redevelopment, reeducation, and reimplementation. Seriously, up until now, everything has been conceptual — who can argue with paper? But you are about to change the business. What can you expect?

It is difficult to give a specific answer to that question, since every organization is different. In general, the implementation goes like this:

- Enthusiastic changing of some policies
- Unbelievably positive improvement

- Less enthusiastic changing of more policies and procedures
- Positive improvement
- Constraint moving to an area not addressed by the initial implementation
- Pretty good improvement
- Leveling off

Most implementations lose momentum after 2 years. The plant is now running like clockwork, costs are down, performance is up. The constraint is no longer in manufacturing. The focus of the business and the improvement projects must now shift to external issues. So, rightfully so, the attention of the organization moves to other areas, not in manufacturing.

However, in those 2 years, your business will change in ways you cannot possibly imagine today. Your performance will level off at the much higher level you are enjoying today. How about a 43% annual ROI? Could you sit there awhile? I know a company that did. How about taking your order fulfillment cycle from 3 weeks to 3 hours and stalling there? I know another company that did that.

The first stage of the implementation will be like housecleaning, with many constraints that you identify and then quickly break. Each time you break one, results improve. This period lasts about 90 days. Eventually, you will find a constraint that will be difficult to break. It might be the market; it might be the product; it might be a $2 million machine.

Then comes the hard work. Implementing the system to exploit and subordinate will take longer than the quick hits you have been getting. If you do not plan for it, the implementation can become bogged down here. This phase may take 30 days; it might take 6 months. It is in this phase that the commitment you have gained in the prior steps will pay off. It is not really that much fun implementing a scheduling process and dealing with people who want to work on product early. You will also encounter here what John Covington calls "the back to Egypt crowd" (as in the Israelites who thought they were better off being slaves in Egypt than being killed at the Red Sea — just before the Red Sea parted). They are the ones who will insist that the business worked better before the constraint management concept came around. They will resist changing. Project deliverables will be missed. People will be "reassigned" because they will not change. It will happen. That is why we implement a measurement system first. If we do not have any measurable results, it will be hard to fight these people.

Summary

The implementation process is a system to build and maintain consensus to change. The technical aspects of the system are straightforward, as you will see. In any event, the particulars will change for every implementation. Therefore, we must address the change management, system design, and project management aspects of the implementation in the process.

The most difficult obstacle is inertia. The implementation process has to move people from working *in* the business to actively working *on* the business. Anything you can do to remove the fear of change will help you achieve your goal.

3 The Implementation Plan

Constructing the implementation plan is the first task of the project manager. The implementation plan consists of real tasks, real procedures, real deliverables, and real dates and is necessary for real accountability. I am not going to teach you how to manage the project; I assume you already know how to do that.

Your plan will be different from this one, but will contain elements in common with this plan. Really, this is the bare bones of an implementation plan. You will probably find other things that need to be done in order to synchronize your business. In other words, use this as a starting point. I will explain each step. You decide what you want to keep and what you want to add.

A detailed plan is a part of getting and maintaining organizational commitment. People need to know where they are going. You will notice that I do not consider a task complete unless there is some form of documentation. I place a high value on documentation for two reasons. One, if you cannot write it down, you do not understand it. The second reason is to speed up the implementation. The policy and procedure documents are your way of teaching the organization how the system works. After the system has been in place, the need for document procedures and policies is diminished. But for now, they are very important.

As you read each item, you will undoubtedly have questions. This chapter is an overview of the steps in the implementation. I want you to have the big picture before I get into specific examples and detail. If you cannot wait, you will find more specific information and examples in the later chapters. Go on and read ahead.

Elements of the Plan

The plan is broken into subcomponents:

- Project launch
- Assessment
- System design
- Production planning and execution
- Organizational alignment

Project launch is when you get the commitment to implement the project. The assessment section is the list of tasks to accomplish to assess your system. System design lists the approach to develop your new system. The planning and execution section describes how to get control of the business. Organizational alignment shows how to get the rest (nonproduction functions) of the organization aligned with the constraint processes and thinking.

Structure

Each step in the implementation plan is broken down into process steps, deliverables, accountability, and acceptance criteria. The process step describes the process being implemented; the deliverable is usually a procedure or policy. Accountability is who is going to make the task happen; the acceptance criteria are how we know the step has been accomplished.

Components of the Implementation Plan

Project Launch

This part of the plan is for the project manager only. The project launch phase is the beginning of the implementation — setting the stage. Before any work is begun, you must achieve commitment and set expectations for the implementation. Done correctly, you will have achieved organizational buy-in to the change process.

The tasks shown in Table 3.1 walk you through the steps required to effectively kick off the implementation. They get the project started on the right foot. The start of the project is when expectations and enthusiasm are at their peak, so you make the most of them, getting the commitment to move ahead before the change work begins.

Table 3.1 Project Launch

Process Step	Deliverable	Who?	Acceptance Criteria
Set objective(s)	Project goals	Project sponsor or team	Written mission statement for the project; launch memo
Determine measurements of the objectives	Project goals	Project sponsor or team	Launch memo
Assign internal project manager and team members	Project announcement	Project team	Letter from project sponsor stating objectives and scope of responsibilities
Reporting mechanism is reviewed	Monthly report	Project leader	Document and project team concurrence
Steering committee named		Project sponsor	Document and project team concurrence
Rough timeline established	Project goals	Project leader	Launch memo
TOC philosophy/ implementation overview presentation		Project leader	Attendee list
Project kickoff		Project team	Kickoff meetings held

Set Objectives

It will be hard to achieve consensus if the project's objectives are poorly articulated. This is the stage where you set the objectives for the implementation. Usually, the project sponsor establishes these objectives. The project sponsor is the senior executive with the organizational and budget authority to authorize the implementation. I prefer the management team (usually reporting to the sponsor) to establish the objectives, but either way is fine.

You should decide what you hope to accomplish with the implementation. The objectives should be measurable and have concrete business benefits. Construct a mission statement for the project.

Determine Measurements of the Objectives

The measurements should focus on the project objectives. This is the first concrete action of the implementation. Do not do anything until you have buy-in to them. The metrics will establish a baseline and set you up for a win. Again, the project sponsor or project team determines the measurements.

Assign Internal Project Manager and Team Members

Decide who will lead the project and who will be on the project team. Typically, the project sponsor will decide who is on the team. Everyone on the team should have a job. The project team is not a place for people to come to meetings and watch.

Reporting Mechanism Is Reviewed

How will results be reported to management and the organization? I suggest you formally report progress on a monthly basis. Some people want written reports; some are satisfied with the measurement report and attending the project meetings. It is a matter of what your project sponsor and management team want.

Steering Committee Named

If your team consists of middle managers, they may want a monthly meeting to review progress. Again, not everyone wants a formal meeting, but depending on the reporting method, a steering committee may be named. The steering committee will be there to help you remove organizational obstacles and get resources to implement the changes.

Rough Timeline Established

At this stage, a time target for implementation should be established. Break the project down into milestones and establish dates for completion. It does not mean they will not change otherwise, but the idea is to establish an estimate of how long the project will last. It should have a definite end.

TOC Philosophy/Implementation Overview Presentation

The presentation of the Theory of Constraints (TOC) is the key to obtaining buy-in to constraint management as a methodology. I like to use a 2-day workshop that shows the main points of CM: definition of constraints, the measurements, the DBR methodology, and the impact on organizational strategy.

Of all the steps I have outlined, this is the most critical. You have to change mind-sets before you change methodology. If people are still thinking in the traditional modes, you will have a very difficult time convincing them of the need to change and they will not understand how each piece of the process fits into the overall system. This presentation also will set the expectations for the implementation — what will happen, to whom, and how.

Project Kickoff

Project kickoff is when the entire organization is notified of the implementation. Most people are not going to be directly affected, but we tell people what we are doing to avoid creating fear in the organization. It is also a step to solidify management commitment. It is very difficult to go back once management has publicly committed itself. Typically, kickoff is an "all hands" meeting or series of smaller departmental meetings.

Assessment

This stage of the implementation is to define the as-is state of the order fulfillment process by interviewing the people involved in the order fulfillment process and determining what is done and why (Table 3.2). Talk to the managers, supervisors, and the people doing the work. My experience is that the managers and supervisors do not really understand the details of what their people are doing. Also, they may not be as forthcoming in their explanations of why certain things are being done. Be sure and go to the source.

Define Order Process Flow

Staple yourself to an order and work through the order fulfillment process from order acquisition to shipping. Diagram the results. Determine how long things take. At the conclusion of this step, you should have a flowchart that shows how orders are processed, by whom, and how long each step takes.

Table 3.2   Assessment

Process Step	Deliverable	Who?	Acceptance Criteria
Define order process flow (order receipt to delivery)	Order fulfillment process	Project leader	Written flowchart
Determine job accountabilities		Project leader	Organizational chart
Identify areas to focus implementation	Implementation strategy	Project leader	To-do list
Identify sources of information		Project leader	Provide location of data or data itself
Define planning process	Planning flow	Project leader	Written flowchart

Determine Job Accountabilities

This may be as simple as getting a copy of the organization chart. What you are after is a document that registers who is responsible for what aspects of the order fulfillment process. This is part of establishing the baseline process. Nothing may change in the new system, but on the other hand, the organizational structure may radically change. Understand where you are before you try to change.

Identify Areas to Focus Implementation

As you walk through the order fulfillment process, you will get an idea of the main things to change. Part of the assessment documentation is this list. You will use this list to create the detailed implementation plan.

Identify Sources of Information

You will have to define the status of order information, bills of material, work in process, part times, and other data elements to derive a schedule. If you are implementing APS software, you will need to identify the location, condition, and accessibility of this information. A manual implementation will require less detail, but you still have to figure out with what data you are going to schedule.

Define Planning Process

In addition to the order fulfillment process, you will have to look at the planning process. How is the plant capacity reconciled to customer demand today? Look at inventory planning and purchased part buffering as well.

System Design

This phase is where you define the to-be state of the order fulfillment and planning processes. At its conclusion, you will have a clear idea of what the new system is going to look like and the primary differences between the vision and the current states. These differences should be clearly articulated with the business benefits and the project objectives in mind. If you cannot see how the new design will have a dramatic impact on the project measurements, you have not done your job (Table 3.3).

Overall Structure of New Order Fulfillment System Developed

Make a flowchart of the new system. Be sure to contrast it with the current system in terms of what work is to be done, who is doing it, and how long it is going to take. You will be using it to get buy-in to the solution.

Policy, Procedure, and Measurement Changes Documented

Prepare a summary document that shows the primary policy, procedure, and measurement changes needed to be implemented. This is the raw material for the detailed implementation plan and is used in conjunction with the previous document to get buy-in to the solution.

Implementation Strategy Developed

You should have some idea of how you are going to implement. Think about the sequence of the implementation. Prerequisites, ease of implementation, payoff for effort expended, and organizational support will all factor into your strategy.

Table 3.3 Design

Process Step	Deliverable	Who?	Acceptance Criteria
Overall structure of new order fulfillment system developed	Vision planning/order fulfillment	Project leader	Document and project team concurrence
Policy, procedure, and measurement changes documented (summary)	Implementation strategy	Project leader	Document
Implementation strategy developed	Implementation strategy	Project leader	Document and project team concurrence
Approvals obtained for change		Project team	Project team concurrence
Approvals of implementation strategy	Assessment	Project team	Project team concurrence
Project measurement system implemented	Metrics listing	Project team	System in place, accountability and procedure documented
Detailed plan	Implementation master plan	Project leader	Document

Approvals Obtained for Change

Take all of the documentation you have prepared to date — assessment of as-is, vision documentation, policy changes, and implementation strategy. Make a presentation to management on the implementation. You must have its support to implement, because the project will cross departmental lines.

Approvals of Implementation Strategy

In addition to the system vision, you have to get buy-in to the implementation methodology. You are not going to get far if the team does not agree with your priorities and methods. One of the obstacles is going to be training. You will be taking people out of their regular jobs for training — only if you have management support. Another obstacle is priority. You may have to explain why you have chosen to do some things early and others later.

Project Measurement System Implemented

Do not get too far into the implementation unless you have the measurements in place. This is the primary tool you will use to report progress. If you do not have any way to report progress, people will question whether you are accomplishing anything. You do not want that to happen. People should be delighted with the project — at all phases.

Detailed Plan

The last step of the design process is a detailed action plan to take you from where you are to where you want to be. The project manager develops the plan, but the project team commits to it. The culmination of this step is a session with the management team. You will probably spend 4 hours explaining the implementation plan and what is in it to the team. Be ready!

Production Planning and Execution

Production planning determines how the plant production will be planned and run. The production planning process is how you will reconcile the demands of the market to the constraint (Table 3.4).

Post-Critical Measurements by Department

Since behavior flows from the measurements, you should decide what the critical measurements are. They will certainly include the project measures, but there will also be others. This step is not completed until the measurements have been defined and accountability is given to gather and report the measures. Post the measures in the work areas. Not everyone will understand what the measurements are or why they are posted, but this will create awareness on the part of managers and supervisors of their accountability. Later, you will use these measurements to evaluate job performance. In the context of the implementation, they show progress and help you diagnose problems in the implementation.

Table 3.4 Production Planning and Execution

Process Step	Deliverable	Who?	Acceptance Criteria
Post critical measurements by department	Critical measurements	Operations manager	Measurements posted in work areas: accountability, procedure documentation
Set the constraint(s)/drum(s), establish capacity targets for the drum	Buffer policy	Scheduling manager	Documentation and concurrence
Buffer sizes and policy are established	Buffer policy	Scheduling manager	Documentation
Develop scheduling policies	Scheduling policy	Scheduling manager	Policy document
Develop scheduling processes	Scheduling process	Scheduling manager	Process document
Develop material/order release policies	Material order release policy	Scheduling manager	Policy document
Develop synchronous manufacturing policies	Synchronous manufacturing policy	Production manager	Policy document
Develop synchronous manufacturing process	Resource allocation process	Production manager	Process document
Specify queue areas and buffer locations on shop floor		Production manager	Areas marked and labeled on shop floor
Generic education of key staff, schedulers, and supervision		Project leader	TOC in production workshop held (2 days)
Shop floor training		Project leader	Attendee list
Communicate new production policies and processes to supervision		Production manager	Training held — "How Rapid Response Works at XYZ Corp."

Task	Deliverable	Responsible	Measure
Department questions and answer sessions held		Project team	Meeting held
Schedulers are trained in new scheduling methodology and processes	Scheduler training presentation	Scheduling manager	Training held — review of new scheduling policy and procedures
Design quick response system (hot list)	Quick response procedure	Scheduling manager	Parameters established, documented, and approved
Implement quick response system		Scheduling manager	System operational, specifications met
Implement "First In-First Out" on shop floor	First-in first-out policy	Production manager	Memo to shop floor
Schedules released to shop floor		Scheduling manager	Schedules being used by shop floor
"Before You Accept the Order" checklist	Accepting orders checklist	Customer service manager	Process documented and implemented
Review inventory policy on standard products and implement appropriate changes (as needed)	Inventory policy and procedure	Inventory control manager	Documentation and concurrence
Change process to open orders	Change order process/change order form	Scheduling manager	Process document and concurrence

(Continued)

Table 3.4 Production Planning and Execution (*Continued*)

Process Step	Deliverable	Who?	Acceptance Criteria
Standard lead-time policy created and implemented	Lead time policy	Scheduling manager	Document and meetings being held with sales, scheduling, and production
Manpower planning policy	Manpower planning policy	Production manager	Policy document
Manpower planning process	Manpower planning process	Production manager	Process document
Daily/weekly capacity review	Action meeting process	Scheduling manager	Process document and concurrence, meeting notes published and distributed
Buffer management implemented	Buffer management procedure buffer analysis	Scheduling manager	Buffer status documented for four consecutive weeks (weekly report generated)
In-depth TOC practitioner training for schedulers (if required)		Project leader	Curriculum documented, completed
Master scheduling process implemented		Scheduling manager	Formal master scheduling review held for three consecutive months

Set the Constraint(s)/Drum(s) and Establish Capacity Targets for the Drum

The first step in the design of the CM system is to determine where the control point is and how you will measure capacity. In many situations, this measurement will not be in hours, but in pieces per hour or dollars per day. At the initial stages, it is not important what the measure of capacity is; just that it is meaningful to the organization and promotes understanding of the status of the constraint. This step may be a lively discussion about what the real capacity of the company is, or it may be nothing more than a memo put out by the production manager. Either way, because the constraint limits the organization, you should publish what that limit is and get agreement from the management team on where the constraint is located.

Establish Buffer Sizes and Policy

Along with determining the constraint, the buffer sizes should be decided upon and published. The buffers have an effect on lead time and inventory. People should know what the factors are in determining these two. It is not a decision to be made by the scheduler alone.

Develop Scheduling Policies

Decide what the role of the scheduler is going to be. What are the limits of his or her authority? What rules are to be used in scheduling the business? How does this fit into the rest of the organization?

Develop Scheduling Processes

What is the exact procedure for scheduling the plant? What data will be used? Is there an approval process? This is what the scheduler uses to do his/her job. We also use it to explain to the rest of the organization how work is scheduled, thereby assuring them of a robust process.

Develop Material/Order Release Policies

The material/order release policies govern how work is released into the plant and when. This policy is the tool to get people to stop releasing work into the plant whenever they feel like it.

Develop Synchronous Manufacturing Policies

The manufacturing policies perform the same function as scheduling policies — to establish the role and mission of the manufacturing department. Their role should not change, but how they fit into the CM system has to be defined and agreed upon.

Develop Synchronous Manufacturing Process

The process being described is not how to make the product, but how to allocate the resources. The purpose is to establish the procedure to assess available and required capacities at nonconstraints, thus efficiently using production personnel resources. This is important because we do not want to have a nonconstraint suddenly become the constraint.

Specify Queue Areas and Buffer Locations on Shop Floor

This sets out a visual monitoring system of the work in process areas. The idea is that you can informally manage the shop by managing the piles of work. If you see work accumulating somewhere besides in front of the constraint, you are alerted to a potential problem. I like to have the shop managed in a visual way. Clear identification of where work belongs and where it does not simplifies managing the flow of work.

Generic Education of Key Staff, Schedulers, and Supervision

When I say generic, I mean "CM-in-production" education. Make sure the fundamental concepts are understood by the decision makers in the business. Typically, that includes front-line management and the supporting staff.

Shop Floor Training

Making the transition from traditional practices to drum–buffer–rope scheduling will mean a dramatic drop in work in process inventory. Most shop floor people associate high work-in-process inventories with job security. When the shop dries up, fear will be created and productivity will suffer. In order to prevent this effect, we need to educate people on the mechanics of drum–buffer–rope. If they know what is coming, they are less likely to resist the change.

Communicate New Production Policies and Processes to Supervision

Ensure that the front-line leadership understands what the system requires. Remember they used to be focused on managing resource efficiency and utilization. Now they are to be focused on delivering product to the buffer on a timely basis — managing the flow.

Hold Department Q&A Sessions

In addition to the conceptual training, we hold department meetings to explain the new measurements, procedures, and practices of the new system. This is to avoid problems and give us maximum chance of successful implementation.

Train Schedulers in New Scheduling Methodology and Processes

The DBR scheduling methodology has to be taught. Your design and the size of the organization will dictate what this training will be. It could be as simple as spending a few hours with the scheduler or it could involve a formal class given to a group.

Develop and Implement Quick Response System

No matter how robust your system is, there will always be customers who demand delivery in less than the current lead time. Recognize it and develop a procedure to shortcut the system.

Implement "First-In, First-Out" on Shop Floor

An important part of subordination is the implementation of "first-in, first-out" at the nonconstraint resources. In the case of a shop that has many jobs arriving in a day, a method has to be devised to determine the sequence in which the jobs arrive. This will dictate the sequence of processing. The procedure addresses two things — one, how to know what arrived first, and then how to make sure people work in that sequence.

Release Schedules to Shop Floor

This is the first major implementation milestone of the project. The new schedules are being used on the shop floor.

"Before You Accept the Order" Checklist

In make-to-order environments, the proper discipline to completely specify the order before production is sometimes lacking. To promote the shortest cycle time possible, the product and order must be completely specified before it is launched into production.

Review Inventory Policy on Standard Products and Implement Appropriate Changes

Depending on your buffering strategy, the stock levels and ordering policies of standard products may be modified. Your implementation may call for stocking certain kinds of products to promote short lead times or stocking product at an intermediate manufacturing step. The objective of the step is to ensure that the inventory levels are reviewed and monitored on a regular basis to maintain consistency with current demand patterns.

Change Process to Open Orders

In-process orders will be changed. Make sure your system design allows for it. Who should be allowed to change an order? Who should be notified? These are the kinds of issues to address.

Create and Implement Standard Lead Time Policy

This policy defines the use of current manufacturing lead times by salesmen as guides to inform customers of current delivery lead times. By promising deliveries based on current backlog, you enhance your ability to keep your commitment and increase customer satisfaction.

Manpower Planning Policy

In many businesses, capacity is determined not by machine availability, but people availability. The objective is to ensure that plant staffing is sufficient to satisfy market expectations (lead times) and maintain maximum return on investment of plant equipment. The planning policy sets forth the guidelines for staffing.

Manpower Planning Process

This spells out the procedure to determine when to add and reduce labor capacity in line with the policy.

Daily/Weekly Capacity Review

The daily or weekly capacity review is done to keep the short-term sales and manufacturing plans synchronized. The process also identifies changes in the market or plant and creates an action plan to respond to those changes.

Implement Buffer Management

This is the second major milestone of the project. This procedure is almost as critical to the long-term success of the implementation as the schedule of the constraint. Buffer management is a procedure to analyze and control the actual and planned content of the buffers and correctly allocate resources. If this is done well, you will maximize throughput, minimize response time, and maximize on-time delivery performance.

In-Depth TOC Practitioner Training for Schedulers

Long term, your success depends on your ability to create and develop people skilled in constraint management concepts. You cannot have CM "done" to your organization; it must come from within. The key to a lasting implementation is to form a deep understanding of CM principles within the organization.

Implement Master Scheduling Process

This process is designed to communicate the global manufacturing strategy to other functions within the organization, provide a formal opportunity to identify potential problems with the strategy, and create consensus to execute the strategy. It is a way to synchronize the long-term plan with the short-term plan.

Organizational Alignment

This step in the implementation is designed to ensure proper subordination throughout the organization. Every function is in a supporting role to either

Table 3.5   Organizational Alignment

Process Step	Deliverable	Who?	Acceptance Criteria
Implement monthly operating report	Monthly operating report	General manager	Report issued, monthly meeting held 4 consecutive months
Pricing policy revised to take advantage of quick response		Sales manager	Policy document and concurrence

delivering orders or delivering product. Therefore, we have to develop policies and procedures that encourage behavior that supports continuous elevation of the system's constraint (Table 3.5).

Implement Monthly Operating Report

The monthly operating report shows the summary of the measurements for the month. Aim for measuring several spots in the business process — quotations, a predictor of order bookings, order bookings, a predictor of plant activity, shipments, and an indicator of whether the business activity is in line with the market activity.

Revise Pricing Policy to Take Advantage of Quick Response

After you have implemented the new process, your lead times will have decreased significantly. This step is to ensure that you use this improvement in the marketplace to make more money. Many times it is a premium charged for shorter deliveries.

Summary

The implementation plan is your guide to keeping the project on track. Invest some time here to make the suggested plan yours. Think through the why of each step. Understand how each step is going to lead toward the project's objectives. You will have to defend each item on the implementation plan, maybe not at the beginning of the project, but certainly later — in the midst of the project. I favor the minimum number of action items to achieve the objectives. Changing the organization is hard enough without doing things that will not create a measurable effect.

4 Project Launch

The Project Launch phase of the project establishes the organizational expectations and clarifies the deliverables of the project. It prepares the organization for change and provides people with an opportunity to challenge the priorities of the decision to move ahead with the project. During the life of the project, the project itself should be the number two priority after serving the customer. During this phase you secure agreement from the organization that this project is the most important thing to do during the next 6 months.

A poorly implemented launch will cause the project to lose momentum before it shows significant results. The project launch process establishes a foundation on which you can build the implementation. It will take at least 90 days before the effects of your efforts become plain. Therefore, the momentum gathered through a good launch must be sufficient to carry you through the period when the critics and skeptics cast doubt on whether the implementation will actually work. In a sense, the project launch process gets you about 6 months of "grace" before any criticism will be taken seriously.

Project Goals and Objectives

The first step is to establish the goals and objectives of the project. Figure 4.1 is an example the document that explains the parameters of the project. This is the "contract" between the project sponsor and the project leader. This document accomplishes several tasks on the implementation plan.

- Sets objectives for the team
- Determines measurement of the objectives
- Establishes a reporting mechanism to management
- Creates an approximate implementation timeline

Memorandum

Date: March 13, 1999
To: Joe Merritt, President
From: Mark Woeppel, Project Manager
Subject: Project objectives and deliverables

Project Mission:
To improve on-time deliveries and reduce customer response time to a point where you lead the industry in these two measures while maintaining product quality objectives.

Subobjectives:
■ Establish a reliable, credible manufacturing execution and planning system that consistently delivers exceptional reliability and response time to customers.
■ Develop skills base of your people.
■ Improve Return on Investment.

Project Measurement:
The effectiveness of the implementation will be measured on the following criteria:
■ Plant Productivity — Throughput dollars shipped divided by plant operating expense. We will consider a 15% increase acceptable.
■ On-Time Delivery Performance — Number of orders shipped on or before committed delivery date (no extra credit for early deliveries) divided by orders shipped. We will consider 95% on-time acceptable.
■ Return on Investment — Net operating profit (before bonus and taxes) divided by book net equity (before bonus and taxes). We will aim to maintain a 50% annual rate.
■ Response Time — Number of days from order entry (start date) to order shipment ("co" date). We consider 3 days for standard product and 10 days for all other product acceptable for order sizes less than $10,000.
■ Defect Free — Percent of orders shipped without defects. Total orders shipped minus returned orders divided by total orders shipped. The acceptable number is 98%.
■ Percent of Company Population Educated — Total population educated divided by total population. Stratified by three classifications: Basic (4-hour introduction), Fundamentals (2- or 4-day workshop), and Practitioner (Fundamentals plus coaching for 3–4 months — our subjective judgment to classify). Our goal is to have 95% educated on the basics; all supervisory, lead people, and above at the fundamental level; and the core implementation team on the practitioner level.

Project Acceptance Criteria
■ System functionality
■ Delivery of education
■ System documentation
■ Measurements benchmark
Joe Merritt will be the judge of the task completion and will accept the documentation.

Estimated Milestone Completion Dates:

Completion Date	Milestone
1/23	Project measurement available
1/30	Assessment complete
2/17	First education
2/24	Top management presentation — 1/2 day
3/15	Procedures and processes defined
4/1	Procedures implemented
4/2	Schedules being used on the shop floor
6/1	Buffer management implementation complete

Project Reporting:
We will report project status to you and the management team verbally as needed and in writing monthly.

Figure 4.1   Project Goals and Objectives

Project Mission

The mission statement is written to clarify the purpose of the project. It is not difficult to find multiple opportunities for improvement, so write the mission statement to prevent going off on a tangent. The mission also sets the tone for the project. Something big is about to happen and it's going to be good!

The mission statement in Figure 4.1 has several components that yours should share. First, the results are tied to the health of the business. The owner of this business wanted to grow market share using response time as a competitive tool. He already had experience with making his customers angry because he could not deliver on time. So, he imposed the credible, reliable execution system as a necessary condition. Second, the mission was related to the financial health of the organization (improve return on investment). He wanted a process that would improve the return on the capital employed. This restricted the ways money was to be invested in the process and how we justified further investment when we saw a need. Third, the objectives were all measurable.

Project Measurement

Each of the mission points has a specific measurement spelled out. If you cannot measure it, do not specify it as an objective. I have also spelled out how each measurement is calculated. This is done so people will know the measures are not arbitrary. There should be no mystery about where the numbers come from or what they mean.

Project Acceptance Criteria

The project acceptance criterion spells out how we know when we are done with the project. Each project should have an objective completion spelled out. This is to prevent the project team from being distracted before the implementation is complete and to keep the project from dragging on forever.

In the example shown, four criteria for acceptance were created: system functionality, delivery of education, system documentation, and measurements benchmark. The system had to work for the business — a subjective determination that the new processes were fully implemented as specified and designed. The education had to be delivered, the people taught. Implicit was that the quality of the education was acceptable. The new processes and procedures had to be documented. The company wanted to be independent of the expert and have at least a baseline of performance. The new processes also had to produce the bottom-line results specified in the objectives. Joe Merritt was the sponsor and judge of performance. He was the defined customer of the system.

Estimated Milestone Completion Dates

Milestone completion dates give direction on how aggressive the implementation is to be. I like implementations to be very aggressive. If you proceed at a leisurely pace, you lose momentum and the project can stall. Using this method also spells out the major milestones of the project.

Project Reporting

This section clarifies how communication is to take place between the project manager and the project sponsor. I am not a fan of writing reports when I can walk in to someone's office and give a 10-minute briefing. If you are updating the implementation plan with completed tasks, that is about all the reporting I think needs to be done (along with the measurements). But, some people like written reports. If you are a consultant, it helps to write a weekly report so people can understand exactly what you are doing in their organization.

Assign Internal Project Manager and Team Members

Naming the project leader and team members is the first indication that you are serious. Obtain a document from the project sponsor that spells out who these people are and what they are to do. This letter should also communicate the level of authority given to the project team and leader. The authority must be spelled out to ensure the organization knows the project team and the leader are acting with senior management's authority. This will encourage people to cooperate (Figure 4.2). The project announcement tells the organization who has the authority to act and that the person will be asking the organization to change.

The mission statement communicates the goal to the organization. The "why" we are doing the project. Of course, we expect wonderful things and the project manager will have the full support and backing of senior management to make the changes needed. Unless senior management is behind the project, it is doomed.

Memorandum

DATE: 7/1/2000
TO: All Hands
FROM: Top Executive
RE: CM Implementation
CC: Project Leader

I'm pleased to announce the formation of a special project team to implement improvements to our operations. This will entail re-engineering our business processes in the operations organization.

John Swanson will lead the project. Mark Woeppel will assist him.

John will invite members to join the team on an as-needed basis. He has my full support to implement the changes required to achieve the mission.

Project Mission:
Unify and align the functional organizations of XYZ Company using a common system (business process and information) to improve operations efficiency while maintaining or improving customer service levels and flexibility for each business within XYZ Co.

You should expect changes to become apparent very soon. My expectation is this will be a positive experience for everyone involved. Please give John your full cooperation and support.

Figure 4.2 Project Announcement

TOC Philosophy/Implementation Overview Presentation

This presentation is to explain to the management team how CM works. When I was a consultant, an outsider to the organization doing the implementing, I was the one to make this presentation. I think it works best to bring in an outside expert to conduct this part of the training. This expert will have the experience in conducting this type of a session, which is not strictly education, but also consensus building. At the conclusion of the presentation, there should be some people who are wildly enthusiastic, some who are very interested, and the rest willing to try it. We are not trying to make champions out of everyone, but to build consensus to move ahead — two different things.

The workshop allows people to understand the theoretical underpinnings of constraint management and allows managers to evaluate how the concepts are used in a variety of business cases. Your workshop should be interactive, using games or computer simulators. This will allow your team to "test drive" the concepts using the cases, so they can contrast the application of constraint management with the way they are at present managing the business. In this way, they will convince themselves that the CM approach will yield superior results.

Summary

Project launch lays the foundation for a successful project. The approach I have shown will help you avoid an unfocused implementation — a project that will not deliver any real business benefit, a never-ending project, or a project that dies prematurely due to lack of commitment or lack of concrete results. You have the foundation to keep the lines of communication open, avoiding misunderstandings of purpose or action. The project team is empowered to implement and is accountable to deliver specific results. Management has oversight. The team cannot get off track unless management does. What are you waiting for? Get busy!

5 Assessment

In this chapter, you will find a variety of real-life examples. I would love to tell you all about company A, but I think it would be more useful to give you a variety of examples from different organizations. My main goal is to teach you the process, but I would like you to understand the examples. Company A is a small manufacturer of decals and signs, privately held, struggling with profitable growth. Company B is a large manufacturer of oilfield products trying to reinvent itself in the face of unprecedented sales growth. Company C is a midsized producer of machined products for the agriculture market working to reduce inventories and improve customer service. All of these were successful in their implementations. The main point here is that although the organizations were very different, the process to assess their business was the same.

What makes up an assessment? I am convinced you have conducted the assessment when you have answered the questions you had when you began the investigation. That is what an assessment is — an investigation into your business.

So, what are the questions? Let us go back to the focusing steps. First, identify the system's constraint(s). Where (or what) is the constraint and how does it affect the different aspects and processes of the business? To *understand* its effect on the business is to derive the cause-and-effect relationships that exist in the business. You are not only going to identify *what* is going on, but *why*. You will come to understand what is driving the behaviors in your organization before you make any changes (well, you don't *have* to, but it will save quite a bit of time). Understanding the constraint is always the focus of the assessment.

How thorough should the assessment be? I hate to give you a consultant's answer, but here it is: It depends. It depends on whether you understand your business well enough to identify the constraint and the policies that prevent you from exploiting it and subordinating everything else to it. You can almost always get a thorough understanding by analyzing the order fulfillment and planning processes. Focus on the order fulfillment process because it has the greater affect on the customer responsiveness and reliability dimensions of service quality (I showed you in Chapter 2 that these are the two most highly valued dimensions). Look at the planning processes because these have the most impact on how resources are deployed. Resources *drive* the order fulfillment process.

How much data should be gathered? Not as much as you might think. You do not need to be a data hound. Most of the time, the data is wrong and will lead you in a wrong direction. You will be most interested in *why* things are done, and that will not show up in the data.

How long should it take? Given a full-time effort, it should not take longer than a couple of weeks to gather the information, synthesize it, and document your conclusions.

Sometimes, the assessment can create unexpected events within and about the organization. Once, I had a manager in tears because the constraint was in her department and she was sure the problems in manufacturing were due to the inept production people and not the drawings her department provided to them. Another assessment resulted in a decision to liquidate the business — before I could present my full report! Yet another report resulted in a major realignment of management. These are the exceptions, of course. Most of the time, things go according to plan.

Assessment Methodology

When the assessment is complete, you will understand the current process and have the constraint and the core problem or policies identified. That understanding will be demonstrated with a process flow diagram of the order fulfillment and planning processes.

You will have an understanding of the measurement systems that exist in the organization. Because measurements drive behavior, you will need to identify them to understand the behavior. You should prepare a summary document to demonstrate your understanding.

Last, you will need to understand the decision processes. How are people making decisions and what data are they using?

Think of these factors — process, measurements, and decision criteria — as the filters you will use when analyzing the business. As you conduct the investigation and understand what is happening in the organization, you will be asking yourself: What is happening? Why is it happening (what determines the behavior of the person doing the task)? What is the process by which this person is making decisions? Using what information? Until you have asked these questions, you cannot really understand what is going on.

These questions are a shortcut to the answer you are seeking. You do not need mountains of data to diagnose your business; these questions will be enough to lead you to the core problem. That is not to say you should not be thorough; always ask more than one person to validate your conclusions. Look for multiple effects from a single cause. Before you go to management with your conclusions, make sure your conclusions have been tested and validated using multiple data points.

The "TOC Thinking Processes" are an excellent tool for analyzing the data you will be gathering. They are the best tool I have seen to learn cause-and-effect thinking and analysis. However, if you have not learned them, you can still come close enough to identifying the core problem to get started in the right direction.

Since the goal of assessment is to identify the constraint and understand its impact, you will have to see if the constraint is internal or external, then identify the type of constraint (policy, resource, or material), then understand its impact on the business (process, people, customer/market). Conducting the assessment is answering a series of questions:

- What is blocking the organization from achieving more throughput?
- Is the constraint internal or external?
- Is there a capacity constraint?
- Is the market the sole constraint?
- Why are sales in the market limited (policy, performance, perception)?
- Is the constraint being fully utilized?
- Does the planning process allow for exploitation of and subordination to the constraint?
- Is there a material constraint?
- Is the material constraint a subordination or policies issue?

Figure 5.1 shows a flowchart of decision making that will guide you through the assessment process.[6]

---

[6] For a thorough treatment on TOC analysis methodology, I suggest you read Eli Schragenheim's *Management Dilemmas.*

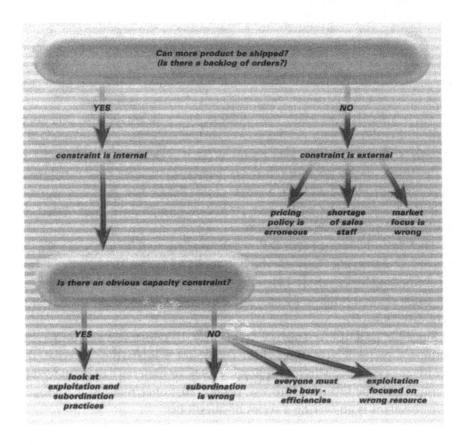

Figure 5.1 Assessment

Interviewing

Conducting the assessment involves a series of interviews with the people doing the work and making the decisions. Always get information firsthand. Conduct these interviews without the department manager/supervisor. The person being interviewed must be free to express what is actually happening without fear of saying the wrong things. I want to stress the importance of confidentiality in these conversations. I have always told my clients and interviewees that the content of sensitive discussions (performance and political issues) will be disclosed, but the source of the information will not. In this kind of atmosphere, you can understand the culture of the organization. You can dig into the processes without people fearing that you are engaged in a performance evaluation (although I am not sure that fear is totally

What is the mission of your function/position?

How does your function/position fit in the overall process?

Who do you work with?

What kinds of issues are you dealing with those people? Why?

What projects are you working on? Why?

How are you measured?   Formal: Informal:

What kinds of problems do you see?

What kinds of changes need to be made?

Is there anything else you would like to share?

Obtain relevant process documentation

Process mapping

Decision points, Criteria for decisions, Quality/Accuracy of information

Figure 5.2   Interview Questionnaire

alleviated). In short, you are more likely to get the truth about what *is* happening and not a story about what *should* happen.

The first person to talk to is the one who has the best view of the overall process. Make a simple diagram of the major steps of the order fulfillment process, who is responsible, and who does the actual transactions. This will yield your first list of interviewees. Take your simple diagram and show it to each person and allow him or her to make changes.

Figure 5.2 is a worksheet to use when interviewing. Not all the questions are geared toward verifying the initial diagram. Engage in a little "fishing." Try to get a sense of what is going on in the organization and a feel for the culture (if you are not part of the organization being studied). This information will help when you construct a strategy for implementation.

The Initial Assessment

The first stage of the assessment is to determine the direction of the subsequent stages. Here you are going to identify the "neighborhood" of the constraint. Another way to determine this is is to ask: "Is the constraint external or internal? Why?" Initially, most constraints are internal and very rarely are they the market or the vendor, because most companies do not dominate the world with the sales of their product and most managers will not tolerate a vendor's being the constraint. Nevertheless, you have to satisfy yourself that these are not the constraints.

The results of this initial inquiry will guide the direction of the assessment inquiry. For example, if the constraint is not internal (there is excess capacity), your inquiry will be focused on sales processes, pricing, sales and marketing policies, and order acquisition processes with a more superficial inquiry into the order fulfillment processes. Of course, when delivery performance (responsiveness and reliability) is the cause of the market constraint, you will focus mostly on order fulfillment processes, with less emphasis on the sales and marketing issues. If you find that you are turning away sales (one sign of a capacity constraint), you will be very interested in identifying the internal constraint and breaking it.

The important thing to remember is that your assessment is investigating why the organization is not achieving the desired performance in the area management has chosen. Your investigation will be centered on what affects those business objectives.

The first step of the initial analysis is to gather measurements of the current performance in these areas: profit, return on net assets, on-time delivery to the customer (reliability), and lead times (responsiveness). These are the measures of financial performance and internal process stability that affect two of the major stakeholders in the business — the owners and the customers. Because you work for the owners, you should focus on their needs first. The customers are second, because satisfying them is the way to attain the satisfaction of the owners. You are not ignoring the other major stake-holders in the business (i.e., vendors and employees), but saving them for later. If you are going to improve the business, establishing a benchmark is essential. You cannot improve what is not measured.[7]

Figure 5.3 shows the current performance indicators of an organization. The goals were to improve manufacturing productivity by 15%, improve on-time delivery performance to 95%, maintain return on equity of 50% annually, and reduce response time to 10 days for signs and 3 days for decals.

The following is a summary of how I conducted analyses of the businesses in which I worked and with whom I consulted. You might think it is in too much detail, but remember that I was always new to the organi-

---

[7] There may be other dimensions you must measure depending on the project objectives.

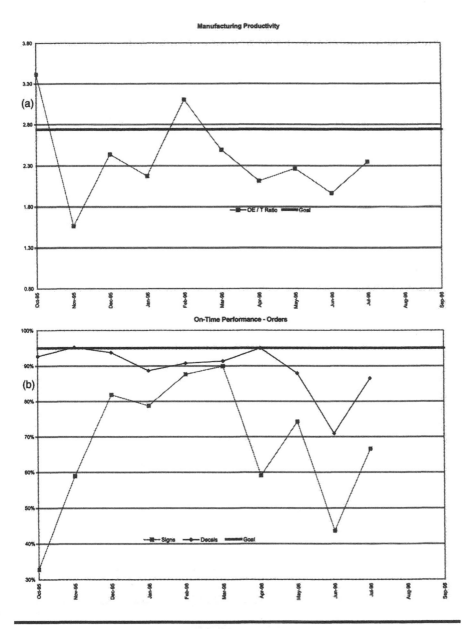

Figure 5.3 Current Measurements

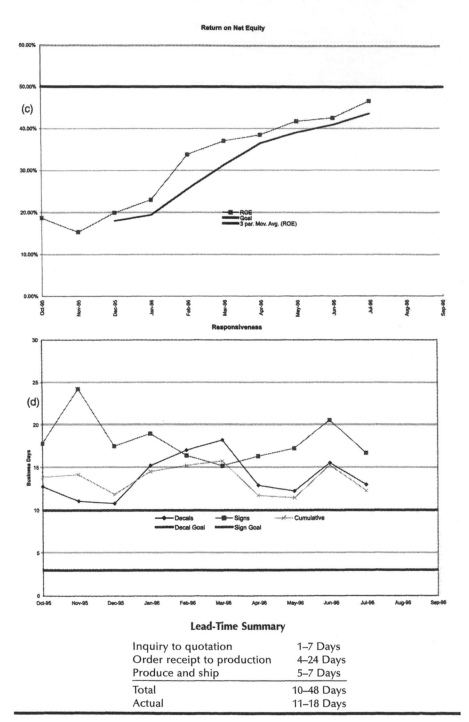

Lead-Time Summary

Inquiry to quotation	1–7 Days
Order receipt to production	4–24 Days
Produce and ship	5–7 Days
Total	10–48 Days
Actual	11–18 Days

Figure 5.3 *Continued*

zation and had to come up with a process to learn the business in a very short time. You probably do not need this much depth — you already have the intuition and understanding. In case you do not, here is the process presented not in chronological order, but broken down by the main process evaluations.

Order Fulfillment Process Evaluation

You are now engaged in a "Herbie Hunt."[8] Analyzing the order fulfillment process is a systematic way to find the constraint. The constraint always shows up as a shortage of resources (finished product, capacity, components, raw material) to meet demand, even if it is a policy — which it is 90% of the time. Evaluating the process is the way of determining which resources are candidates for the constraint.

Figure 5.4 is a diagram of an order fulfillment process. The point of producing the diagram is to identify problems in the global process that need to be addressed in order to exploit and subordinate the constraint. In the case of the process shown in Figure 5.4, the major problem was that order promising was not connected to the production plan. You cannot derive that from the diagram, but there is a hint of it if you will look at when the order actually went to scheduling.

In the process shown, the implications for subordination and exploitation had to do with the release of orders to production before the products were fully specified and designed. As you can see, most of the processes and decisions were focused on understanding and translating the customer requirements into manufacturing requirements. The process design had to provide a way to reserve capacity in manufacturing before it was fully understood what would be required. Customers demanded delivery information early in the sales process, although it was not fully understood what, exactly, the product specification would be. Before orders were released to manufacturing, the product had to be fully specified to avoid delays in production and wasted time on the constraint resource.

Do not get caught up on the specifics of this diagram; it is only an example. The point is to produce a document that shows the order fulfillment process in a way that the management team can understand.

---

[8] The term "Herbie Hunt" refers to the Boy Scout hike analogy in Goldratt's *The Goal*. Herbie was the slowest Boy Scout on the hike — the bottleneck or constraint.

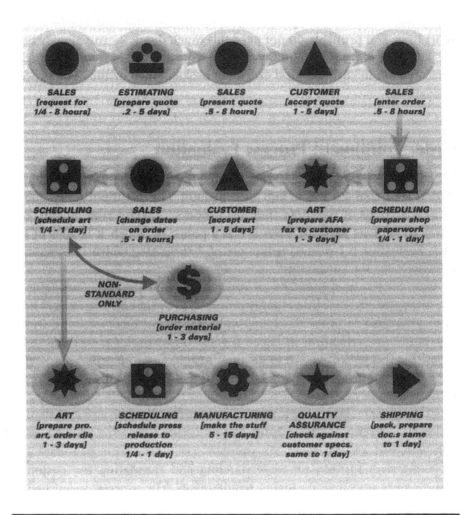

Figure 5.4   Order Fulfillment Process

Planning Process Evaluation

Figure 5.5 is a diagram of a planning process. The goal of diagramming the process is to ensure that you understand all the needs of the organization when you develop the new process that centers on exploiting and subordinating to the constraint. You will understand why you are not getting the results you want. In the case of Figure 5.5, the main problem was that purchasing decisions were not connected to production activity. Materials were bought from the MRP run, but the MRP did not reflect what was being done in production.

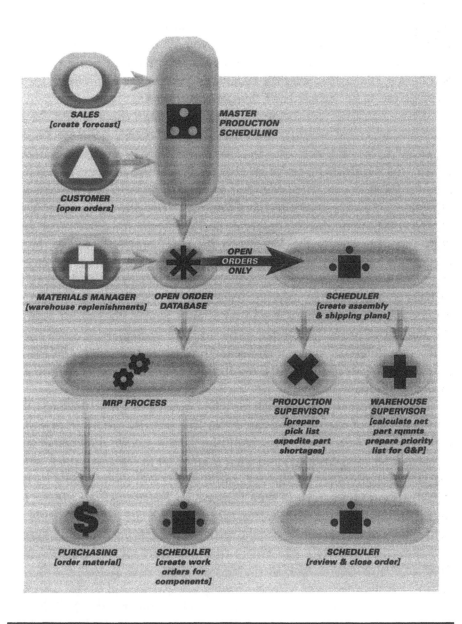

Figure 5.5 Planning Process

When you analyze the planning process you will understand how the organization reconciles customer demand (which varies) to resource availability (which also varies). In Figure 5.5, you see how demand is placed on the resources by customer orders and a forecast. In a plant using a forecast, you have to understand how the forecast is created. Most important are the assumptions about demand and demand aggregation.

When evaluating forecasting techniques, the main thing to look for is a phenomenon called *overshoot*. Forecast overshoot occurs when the forecast is padded in multiple iterations to cover uncertainty in multiple, sequential links in the supply chain. Let us say that retailers create forecasts of end item sales for the next year. They give their forecast to regional distributors, who aggregate and synthesize the retailers' forecasts. They give the forecasts to the company salesmen, who also aggregate and synthesize the distributors' forecasts and then present the forecast to the plant master scheduler. In the companies I have seen using a forecast, the main complaint is the lack of product. Most distributors or salespeople are not measured on inventory (or forecast accuracy) but on revenue. Their thinking (rightfully so) is that you cannot sell out of an empty wagon, so at each step of the forecast creation the amount of product projected to be sold will be higher (to cover uncertainty), resulting in a higher than needed product requirement. This has the potential to overload the facility, resulting in additional capital spending or hiring of too many people.

There is another effect to look for when making to stock (forecast). The same problem of overordering can distort the priority system in manufacturing (Figure 5.6). Often, the scheduling system cannot distinguish what is a real order for a customer and what is an order to replenish stock. When the plant resources are overloaded, some orders will undoubtedly be late. The distribution centers will then respond by *increasing* their orders, further flooding the plant with orders, creating more late orders, causing the distributors to increase their orders, until manufacturing begins an aggressive campaign to reduce late orders through overtime and outsourcing.[9] Your goal is to ensure that, in the design of the new system, the factory planners will understand what is actually being delivered to the customer (i.e., distinguish between replenishment and customer orders). Only then can the factory planners make intelligent decisions about when to add or reduce capacity.

---

[9]   The solution to this problem is outlined in the book *It's Not Luck* by Eli Goldratt and is also described by Peter Senge in *The 5ᵗʰ Discipline* in his discussion of The Beer Game.

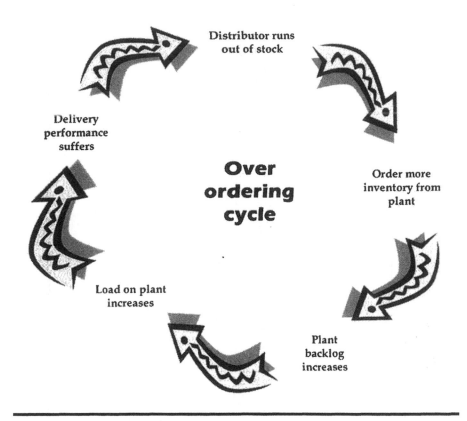

Distributor runs
out of stock

Delivery
performance
suffers

**Over
ordering
cycle**

Order more
inventory from
plant

Load on plant
increases

Plant
backlog
increases

Figure 5.6   Overordering Cycle

Determine Job Accountabilities

Determining accountabilities can be as simple as getting a copy of the organization chart (Figure 5.7). What you are seeking is a document that registers
who is responsible for what aspects of the order fulfillment process. This is
part of determining the baseline process. Nothing may change in the new
process, but on the other hand, the organizational structure may radically
change. Understand where you are before you try to change.

Identify Sources of Information

You must define the status of order information, bills of material, work in
process, part times, and other data elements to derive a schedule. If you are
implementing advanced planning and scheduling (APS) software, you will
need to identify the location, condition, and accessibility of this information.

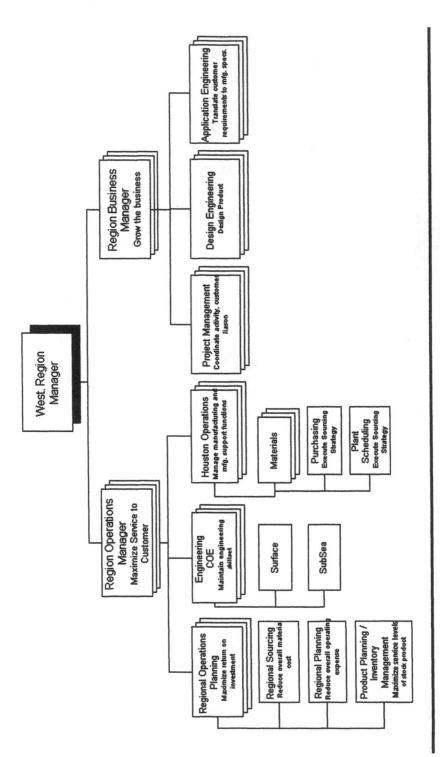

Figure 5.7 Job Accountabilities

Action Items
- Philosophy of pricing document
- Gains to derive 1996 goals
- Monthly business planning — volume vs. capacity
- One person dedicated to manage large products and customers
- Special/different handling on reruns or small orders?
- Analysis of throughput/capacity constraint resource on product lines
- Quotes contingent on final approval and art
- Stratify prices based on deliveries
- Focus on image to customers (physical, occasions, telephone)
- Employee survey

Figure 5.8   Action Item List

A manual implementation will require less detail, but you still have to figure out with what data you are going to schedule the constraint.

Identify Areas to Focus Implementation

As you analyze the order fulfillment and planning processes, you will identify the system's constraint and the prerequisites to fully exploit and subordinate to it. As you progress, you will make a list of the main elements required in the final system design (Figure 5.8). Part of the assessment documentation is this list. You will use it as a springboard to create the detailed design and implementation plan.

Summary

In conducting the assessment, you will be directed to the location of the constraint. You will also gain valuable insight on possibilities to break the constraint or fully exploit it.

When conducting the assessment, you will find many things that need to be fixed. Remember to stay focused on the constraint. You will find more problems and opportunities that require "immediate" attention. Remember that there will always be things to fix and not all the problems you encounter during the assessment are central to the issues of the constraint (i.e., identification, exploitation, subordination, or elevation). Some of the problems that you find will be only irritants. For example, you may find that warehouse racking is inefficient and implementing a new picking process would reduce manpower requirements in the warehouse. If the constraint is in the plant, who cares? There is far more impact to be made increasing the output of the constraint than there is reducing warehouse staff.

The level of detail you will gather will depend on the level of your understanding of the organization and your ability to communicate what is happening to the organization. Remember, the goal is not to do analysis, the goal is to change your organization. However, you may do a detailed analysis or assessment of the organization just to convince the other members of your team that such an assessment has been conducted so that they will buy into your solution.

The assessment documents will be one weapon to use in the battle to change the organization. As-is process diagrams and current performance measurements can show the system really *is* broken. Unless people believe it is broken, they will not want to fix it.

6 Designing the New Process

The step of designing the system provides a map of what the new process(es) will look like and how all the elements work together. By defining in specific terms how the end will look, you can engage the organization in a (mostly) positive discussion of what change should be implemented. The design procedure has two distinct components: development of the new process itself and the positioning of the solution to get management buy-in.

Common themes apply to all TOC-based process designs; the constraint is determined ahead of time (identify the system's constraint), a process exists to provide a mechanism to exploit the constraint, a subordination process is developed, and information is provided to identify the location of the next constraint (elevate the system's constraint). Regardless of anything else you may implement, the new business processes will include a drum–buffer–rope scheduling system to exploit and subordinate the constraint, and you will have a system implemented to monitor the status of the buffers and report statistical data on them (to identify the location of future constraints). This process will be the basis of the design. Of course, during the assessment process you will find duplication of effort, inaccurate or nonexistent communication between functions, unnecessary delays, and other problems. You might as well fix them while you are developing the new process.

You could implement many different solutions. Some fit your situation, and some do not. The big idea behind constraint management is not that there is a silver bullet to fix all your problems, but there is a framework, which allows you to effectively analyze and make decisions, that is fundamentally different and better than the prevalent cost-based paradigm. Therefore, I am not going to explain all the solutions, but I will give you some ideas to think about when developing your new processes.

The Design Process

The design process begins with choosing the constraint. It probably will not be where it is now. Usually, that is the case, since you are not getting the results you want. Next, identify the new processes needed to exploit that constraint and subordinate and synchronize the rest of the organization to those processes. Create a process map of how the processes will work together. Identify the changes from the current practices and develop a strategy to move from where you are today to where you want to be. Last, document and present your solution to management.

That is all there is to it! Sounds simple, doesn't it? This is where the knowledge of TOC comes in — knowing which solution applies and how it fits with your organization. I hope the last half of this book will help in the specifics.

Choosing the Constraint

Your first decision is to choose your constraint. Since we will always have a constraint, your first decision concerns where to place the constraint and control points in the business. You are going to invest significant effort and time into developing and implementing new business processes. These processes will be designed around the constraint and control points. The operations and sales strategy will revolve around the location of the constraint. You do not want to redesign your business every time the constraint moves, unless you like to redo your work every few months.

The choice of the constraint is guided by the strategic intent of management. From there, tactical policies are derived to guide the organization to exploit and subordinate to the constraint. Out of the policies come the detailed procedures. I think the most important part of the design is creating alignment within the business. Therefore, you cannot ignore the organization's strategy in developing the design. If the design is out of alignment with strategy, you will not get buy-in to implement; management will not allow it.

Implications of the Constraint

You cannot escape the effect of the constraint on your business. As the weakest link, the constraint will determine the overall profitability of the organization. Remember that the worst performer in your system dictates your overall performance. This is not simply a matter of choosing which machine will be the bottleneck. Your overall financial performance, including return on investment (ROI), is determined by how well you exploit the constraint of the system (see Figure 6.1).

Constraint vs. Control Point

The control point is the area or resource that, by close management, enables you to effectively manage the order fulfillment process. It could be the constraint resource, but not necessarily. It could be a bottleneck resource. The control point is the *drum* for your plant.

Do not confuse your decision of where the constraint should be with where the control point should be in your scheduling system. Although they are related, they are distinct. The constraint is what limits the organization from making more money. The control point determines the rate of production and subordination efforts. Your planning and scheduling system will revolve around the control point. In effect, the control point is either the constraint itself or a shadow constraint. The *shadow constraint* is a resource that is subordinate to the constraint but mimics the constraint's effect on the internal operations of the business.

Your control point choices are either the market or an internal resource. Keeping the control point internal allows you to smooth the variation in market demand, which in turn allows for a more efficient use of resources. However, this choice requires additional buffering either in lead times or inventory to absorb those variations, thereby increasing your investment requirements. Putting the control point in the market keeps your lead times low in times of nonpeak demand, but requires higher operating expenses to maintain protective capacity to absorb peaks of demand. Having the order delivery requirement be the control point is no guarantee of short lead times — especially when the peak demand arrives.

Your choice is similar to that of a restaurant trying to control the waiting time for tables. The number of meals a restaurant can serve is usually limited to the number of tables they have. Keeping patrons at the table for long periods reduces the amount of food that can be served. So to improve profitability, one can "turnover" the tables multiple times during meal times, thus

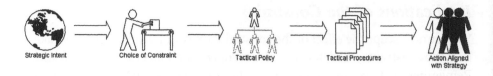

Figure 6.1   Action Aligned with Intent

increasing the amount of food sold and revenue generated. Maintaining the control point internally would dictate preparing food ahead of time. This would provide maximum utilization of the constraint (tables) to ensure the kitchen does not block utilization of the tables by increasing waiting times for food. This strategy maximizes constraint resource utilization, but limits the choice and freshness of menu items, potentially limiting customer choice (and thus limiting potential markets). There are also losses (increased operating expense) associated with food that is prepared ahead of time, but not sold. Maintaining the control point in the market would dictate the preparation of meals when the customer wants it, maximizing freshness and minimizing food loss, but in times of peak demand, waiting time increases. This is why understanding strategic intent is vital. Customers who want a leisurely meal freshly prepared will generally pay more than those customers in a hurry.

Your choice of the control point vs. the constraint will probably not be at one extreme or another, but a combination of the two, maximizing speed on some products, minimizing material loss on others.

Financial Performance

Return on investment is a primary consideration when selecting of the constraint. Cash on hand represents opportunity — opportunity to invest in new markets, new equipment, new businesses, modernization, growth. This is rooted in the idea that the marketplace is essentially limitless. The only thing that ultimately limits the organization's ability to grow is the organization's definition of its market. Like a plant, a business is always growing (or at least evolving). Also like a plant, if the business is not growing, it is dying. In order to foster growth you must continue to refine your definition of what and who your market is.

Maximizing profits and ROI, you are competing for money — maybe with other customers at your bank, maybe with other divisions in your corporation. Your competition for money may involve operating expenses — after all, if you do not have cash, you are borrowing, which will have an impact on the bottom line in the form of interest expense. Lack of cash has an impact on growth as well, because cash represents opportunity for new products, markets, and technology. Therefore, your efforts to improve return on investment should focus on eliminating whatever else you might be thinking of using that money for.

This is the first criterion to examine when looking to place the constraint — the effect on return on investment. In order to do this, ask these questions: Where is your capital investment pooled? and What part of your business represents the single largest investment of capital? The actual amount is not significant, only its relationship to the other investments in other areas of the business. For example, in one company I worked with, they have a collection of horizontal boring mills that cost $2,000,000 apiece. One of these will build a new plant! In another company, they have a production line investment of $150,000. This area also happens to be the focus of the labor intensity of the business. You see the extremes: both represented significant capital investments for the businesses in which they were. By exploiting your largest capital pool, you will increase your ROI by maximizing its effectiveness.

There are other questions to ask: How easily duplicated are these resources? and Are they difficult to outsource or to locate vendors to flex capacity up and down? My experience has been that these kinds of resources represent the core competency of the business and as such cannot be easily sent to outside vendors. These are good candidates for control points or constraints, because in making them the constraint, you will focus your efforts on maximizing your strengths while simultaneously subordinating the less important areas to that strength.

Market Responsiveness

It does not help the business much if you maximize the resources, but have poor responsiveness to the customer. Ideally, you maximize your investment while simultaneously eliminating your competition — exceeding the needs of your customers on delivery performance and lead times.

In one company, the desire to maximize the investment was primary, but they could never get product to the customer. They had almost a month's worth of production past due, but they were efficient! In effect, they were saying to their customers, "Sorry, we can't deliver right now; we're being efficient." In another company, they deliberately suboptimized the investment to maximize service (to get sales growth). This was done because, ultimately, the constraint was in the market (as it always is) and it did not matter how good they were in the plant if they did not have product in the warehouse to sell.

Market responsiveness is another consideration in locating the constraint (and control point). If, your constraint is in the market, weigh your desire for profitability today against your desire for customer satisfaction tomorrow (and potentially your profitability tomorrow).

Growth Strategy

Closely related to market responsiveness in its implication for constraint selection is the overall growth strategy of the business. Do you plan to be the low-cost provider? If so, you will give more weight to maximizing internal resources. Are you growing through superior service and customer intimacy? In this case, you will give less weight to maximizing internal resources and more weight to process flexibility and customer service levels. Are you the smallest player in the market with a plan to be a major player? Then, you will be more inclined to suboptimize current returns for increased market share.

The point is you cannot ignore the overall business strategy in choosing the constraint. It may be that the constraint will change the strategy, but it will be more likely that the macro forces of your business strategy will dictate the choice of the constraint.

Workforce Strategies

The choice of the constraint will affect your workforce strategies. The most common and obvious choice is the decision of staffing the constraint resource. Since this resource has the largest effect on the performance of the business, you would naturally want to have the highest skilled workers manning this resource. As mentioned previously, in one company, the constraint resource suffered from high employee turnover. When management personnel looked at the compensation plan, they discovered this resource paid less than *any* position in the plant. They immediately made this skilled position the highest paid in the plant and were able to attract and keep the most proficient operators. Schedule conformance improved, throughput went up, and so did profits.

Pay is not the only aspect limiting performance. In another implementation, management found the operators were limited by their understanding of the task. They were producing a great deal of rework and scrap. They

Product	A	B	C
Sales Price	$180	$240	$180
Raw Material Cost	$65	$95	$65
Throughput	$115	$145	$115
Time at Constraint	0 min.	34 min.	14 min.
Throughput per Minute	∞	$4.26/min.	$8.21/min.

Figure 6.2 Products with Different Rates of Constraint Resource Consumption

focused training on these people and scrap and rework dropped dramatically. An increase in overall business performance followed immediately due to the increased production at these resources.

As a general principle, your best, most highly trained, most highly motivated workers should be working on the constraint resource. In the case of a company constrained by the market, these might be the salespeople or engineers. Your new process and implementation plan must consider the workforce skills and attitudes as part of the exploitation process.

Pricing

The location of the constraint will have an effect on the pricing practices of your business. In the case of a business that is constrained by an internal resource, those products that cross that constraint should be evaluated in light of their contribution to overall throughput (see Figure 6.2). Those products that have a low contribution vs. consumption ratio are candidates for price increases. Those that have high contribution ratios are candidates for aggressive sales and marketing activities — including potential price reductions. Those products not consuming constraint resources are candidates for opportunistic pricing or adjustments to lower the selling price to stimulate demand and consume excess capacity.

Sales Strategies

The determination of the constraint may trigger additional sales activities to move the constraint from where it currently is to where you have decided you want it to be. As I said earlier, you would also be pushing the sales organization to sell more of the higher contribution ratio products and those that are on nonconstraint activities.

Promotions

Promotion activity could be geared to reducing the seasonality of demand on the constraint resource. You may initiate promotions on products that do not cross the constraint and thus consume excess capacity. Promotions have the benefit of being short term, and thus do not commit you to the risk of a permanent price reduction. However, they have the drawback of creating peak demands on manufacturing resources, which increases the risk of reducing delivery performance. Your decision on the location of the constraint and where it is now vs. where you want it to be should influence your product promotion efforts.

Distribution

Your distribution strategy will also be influenced by the location of the constraint. If you have chosen the customer-intimate, high service strategy, you may want to implement a replenishment/pull system to serve the market. Your buffering strategy will influence where inventory is located and the amount. For example, you may implement the TOC distribution replenishment system, with consignment inventories at retailers, regional inventories to serve those retailers, and a central warehouse to serve the regions. But if your products are make to order, this method would not make sense.

Wrapping Up the Design

The objective is to implement a solution that fits your business. That fit will be different at different times. Therefore, do not aim for an "optimal" design. Your efforts should be concentrated on a process design that has a high likelihood of being implemented, and, once implemented, will work.

The location of the constraint today and your decision on where you want it to be will have a significant effect on the final process design. The decision you make has implications for every aspect of your business and its competitive strategy. If you do not take the time to evaluate these decisions, your process design will be flawed or have a limited life. If your goal is to have a lasting impact on the profitability and competitive position, you cannot ignore these aspects.

During the design phase, the most important decision will be the location of the constraint and the control points. All other design considerations must follow these for the system to produce superior results. This is another reason education on constraint management concepts for management is so important. You cannot choose the constraint if you do not know what a constraint is.

Achieving Buy-In to Implement

Assuming you have done a proper analysis and design, getting the new processes implemented is a task requiring change management skills. Since you have already done your homework, gotten buy-in to begin the project, and completed the initial education, the management team will be receptive[10] to the new process design and your strategy for implementation. You will have your assessment report completed. The new design will be presented showing the new process flow — order fulfillment and planning processes. There may also be additional changes to simplify the process (e.g., automation, transaction timing changes) that are not considered specific to the constraint management concept, but are essential to proper exploitation of and subordination to the constraint.

Documentation

Providing documentation of the future state is critical to getting people to buy in. The presence of documentation demonstrates you have carefully examined the current system, thought through the implications of changing it, and have a plan for making the transition from where the organization is today to where it will be in the future.

I cannot emphasize enough the importance of having this document. It has been said that the crews of Christopher Columbus' ships were ready to mutiny because they never thought they would reach land. Columbus had no map, just an idea. Are your people like the crews of Columbus, ready to go on an exciting journey with an uncertain outcome? I did not think so. Read on.

Process

Figure 6.3 is the vision for the process analyzed in Figure 5.4. What, exactly, the process is doing is not as important as the communication of how the new process will look. At this stage of the implementation, you are dealing

---

[10] Receptiveness does not mean consensus. At this stage of the implementation, your job is to get agreement to move ahead. Be prepared to modify your solution in order to get consensus.

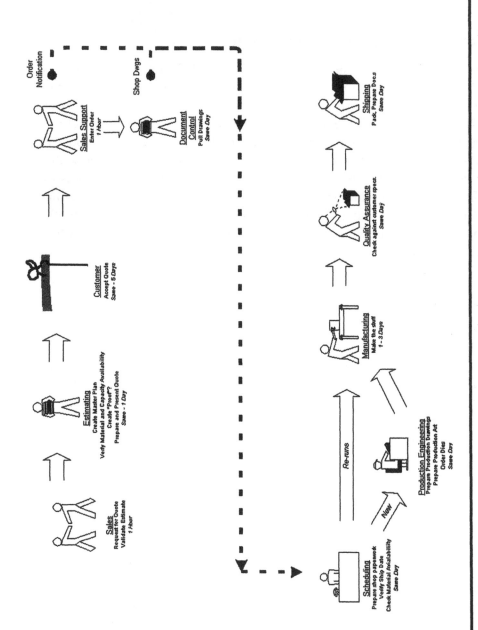

Figure 6.3 Vision Order Fulfillment Process

with concepts and ideas. Therefore, you do not need to fully specify all the procedures and steps. What you do have to specify are the major differences and improvements.

In the example shown, the constraint was in the market. The key to capturing additional market share was lead time and improved reliability of delivery performance. If you will look back at Figure 5.4, you will notice there are 13 steps before the product is ready for manufacturing. The new process in Figure 6.3 has 7. In addition, the new process stops in the scheduling process only once, instead of three times, which allows for more accurate delivery promises.

Figure 6.4 is the "to-be" or "vision" process to replace the one shown in Figure 5.5. This process is from a different business than the order fulfillment process in Figure 6.3. The constraint in the first business was the market, but they had seasonal demand, so accurate forecasting and scheduling resources ahead of peak demand in summer months was essential. The constraint in the second example was in the existing planning process. Purchasing activity was disconnected from production activity. So, the most important design change in the new process was to ensure that purchasing was done based on production activity, not on the master schedule (forecast). The new process also uses a computerized scheduling system to generate net requirements and work orders, a process that was previously done manually.

Your objective is to produce a *vision* of where you are going to go with the implementation, not a detailed design. The process could be compared with architects winning a contract to design a building. While they do not design it in detail, they do construct a model of the building with the building's main features so they can explain the design. The detailed designs — structural, electrical, plumbing, etc. — are worked out after the overall design is approved. In the same way, the policies and procedures to support your vision process design will be worked out in the implementation process.

Policy and Procedure Changes

The policy and procedure changes document is a written summary of the features and changes that will be required to implement the vision process. The addition of text to the diagrams allows for more thought and discussion of the issues for implementation. Again, the list does not have to be all encompassing, but if you are thinking about reorganization or changing the way the organization does its work, include that information. It is at this time that you will have the most spirited discussions and encounter the most

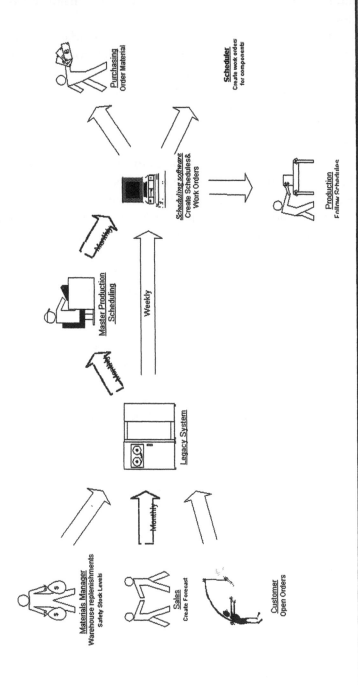

Figure 6.4 Vision Planning Process

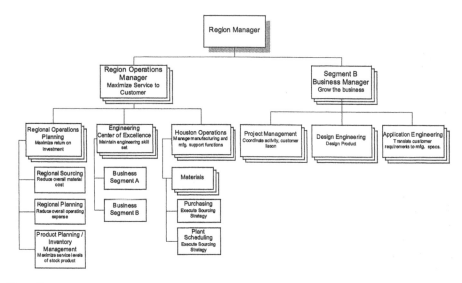

New Processes

Electronic order action submission by salesmen to reduce load on engineering resources

Field service and applications have ability to view production schedules and reserve capacity and parts real time to eliminate calls to the project manager in engineering

Planning bills and routings concept to allow manufacturing and engineering to work concurrently

Product definition standards to ensure no work is done on orders that are not fully specified.

Creation of distribution centers to put invnetory close to the customer and improve responsiveness

Creation of regional sourcing and planning functions to eliminate duplication of effort

Creation of application specialist function to allow separation of engineering effort from sales effort

Electronic order reservation and due date quotation to prevent overcommitment of plant resources

Creation of formal conflict resolution process to limit order expeidting to those really needing it

Creation of a 3-date system to track customer request, promised delivery, and scheduled delivery

Figure 6.5a Vision Process Changes

resistance, because you are no longer talking theory. Now you are targeting individual sacred cows to be butchered.

The companies described in Figures 6.4 and 6.5 were wasting constraint resources, missing delivery dates, and losing sales. Organizational changes were recommended to provide proper subordination to the constraint and

Deleted Processes

- Material and capacity analysis by field, scheduling, planning
- Application engineering in the field
- Planner function
- Using ERP system to promise deliveries and assess resource availability (infinite capacity, no reservations)
- Creation of sales orders for forecast demand on stock product
- Order action in purchasing prior to capacity validation and project scheduling

Revised Processes

- Measurement system revamped to emphasize global performance
- Creation of process to discern product/segment profitability
- Change reporting structure of engineering group
- Creation of dedicated business unit project management group
- Bolstering preproduction planning process
- Three-stage production planning — rough-cut validation at order entry, second allocation at regional planning, final cut at plant scheduling
- Production definition done prior to manufacturing order release — implement go/no go
- Sales and field personnel have online access to manufacturing execution and material plans
- A product planning approach to inventory management
- Purchasing activity takes place *after* capacity planning
- Creation of a plant manufacturing engineering database
- Converting quote to order using ERP system

Expected Results

- Shorter response times to customers (better planning of capacity, fewer surprises)
- Stock product — 84% reduction
- Engineered product — 10% reduction
- On-time due date performance greater than 90%
- Engineering productivity increase of 20% (better priority management, better product definition)
- Organizational alignment within region
- Reduced rework and expediting expense
- Greater manufacturing productivity (more consistent priorities and fewer stops and starts)
- Greater purchasing productivity (more time to source correctly, better product definition)
- Better responsiveness on customer changes (focal point for execution)

Implementation Issues

- Implementation of measurement system requires infrastructure changes and a new mindset for managers
- Skills requirements and availability are not well defined
- Changing the existing bills of materials is a large task

Continued

Figure 6.5b **"To-Be" Process Change Summary**

Implementation Issues

- The "how to" transition from the current process to the new process has not been defined
- Product development lead time has not been addressed
- Accountability for promise dates is weak — will require enforcement action
- How to manage project status is not defined
- No policy exists to define raw material inventory as it relates to engineered product and expected market demand.

Engineering Issues

- Ease of identifying the longest lead time raw material for synchronization purposes
- Capacity management issues
- Engineering changes' effect on promise dates
- Currently, no planning routings exist
- Bills and routings do not reflect the way things are done in the plan

Figure 6.5b

continued sales growth. As in one of the previous businesses mentioned, the order fulfillment process was just too long because of poor subordination of engineering resources to the customers.[11] In this business, manufacturing was subordinate to engineering, so by addressing the engineering issues, they could successfully synchronize themselves to the market.

When you are producing these documents, demonstrate the benefits of the new processes — shorter lead times, better utilization of the constraint, etc. Remember to express the benefits in the context of the project objectives.

Implementation Strategy

The implementation strategy document is available for management to scrutinize and make suggestions concerning the direction and priorities of the implementation. Think of it as a 40,000-foot view of how the vision process will be implemented. It is the skeleton of your implementation plan.

The implementation strategy in Figure 6.6 was geared (1) to provide better input into the engineering function to manage the sales orders (exploiting the desired, future constraint); (2) to break out products that did not need engineering attention to bypass the bottleneck in engineering; (3) to implement the center of excellence concept (pooling engineering resources to

---

[11] The constraint was a single engineer who designed (configured) the product for the customer. This was not recognized by the organization, and thus its efforts for subordinating were misdirected. It was focusing all its attention on the plant.

Implementation Goals

Consistently deliver customer orders on time

Reduce delivery time of customer demand

Implementation Sequence (Delivery Reliability, then responsiveness)

1. Build Consensus for Change
   Present Order Fulfillment Process to key management — adjust and modify
   Present Order Fulfillment Process to remaining involved management — adjust and modify

2. Develop Implementation Project Plan

3. Sales order management (buffer management) in final assembly

4. Creation of formal conflict resolution (escalation) process

5. Implement the Stock Product/Distribution Center process
   ■ Define the following: stock, nonstock, standard, nonstandard, and engineered product
   ■ Inventory locations
   ■ Inventory policy and accountability
   ■ Electronic ordering and reordering

6. Creation of Engineering Center of Excellence

7. Creation of Application Specialist Function
   ■ Planning bills and routings
   ■ Product definition standards
   ■ Product definition done prior to manufacturing order release
   ■ Creation of independent engineering information database

8. Creation of regional sourcing and planning functions
   ■ Creation of product planning focus in inventory management/region planning
   ■ Creation of process to discern product/segment profitability

9. Implement the order promising/reservation process
   ■ Field sales and applications have ability to view production schedules real time
   ■ Creation of 3-date system
   ■ Automated order reservation and due date quotation

10. Reconciliation of quotation to actual performance process

Implementation Process

1. Refine implementation project plan

2. Establish the process implementation sequence with required completion dates

3. Define and document necessary policies, procedures, and job descriptions

4. Implement measurements

5. Define skill sets required and compare against available

6. Acquire skill sets

7. Provide training to all involved personnel

8. Implement plan

9. Monitor performance and modify as required

Figure 6.6   Implementation Strategy

allocate as projects dictated); and (4) to change reporting structures (subordination) to create organizational alignment.

This company is a large, multinational manufacturer. The dynamics of the market had changed significantly over the previous 10 years. The industry had gone through a boom in the late 1970s, a bust in the 1980s, and a resurgence in the 1990s. During the previous 10 years, the customer and competitor landscape had changed dramatically, requiring a different approach to the market. The sales organization was coping as best it could, but the people in the plant were having a difficult time coping with the new demands being placed on them. Many were long-term employees accustomed to doing things a certain way, and new leadership had been brought in to "turn things around."

The implementation strategy focused first on creating a reliable process. But even before that, there was a consensus building process to explain the need for and prepare for change. The strategy was to work on infrastructure development in the plant, then as those efforts matured, to move out to the market with distribution, and finally to build the other internal functions. The idea was to first fix the things that were broken, then to work on refinements.

When developing your strategy, the first consideration is the people to be changed. How can you best get their *active* support for the implementation? Having the people who will ultimately be responsible for the processes perform the implementation tasks will ensure continuing support throughout the project.

Approvals

Before you launch the implementation, you must have the approval of the management team who will be accountable for the results. You cannot assume that just because the senior executive gives permission to proceed, you will be successful. Your implementation will be successful because the people doing the work will be the ones developing and implementing the new policies and procedures. Therefore, you must have buy-in from them. One way to achieve this is to take your assessment, vision, process changes summary, expected benefits, and implementation strategy to the management team for comment. I suggest you take your draft of the report to the key leaders/influencers on the team for comment. Let them tell you where the minefields are, where the sensitive points are. Make changes as suggested. These suggestions will be mostly points of emphasis and deemphasis and will help you get your plan approved.

Unless you have some sort of sign off at this stage, you are taking a risk of having your project undermined. You will not even know it until you are trying to get people to make changes and you discover the conflicting signals — change/don't change. By that time, you will have to overcome the negative bias, not the neutral bias some members of management had at the start of the project — a much more difficult task.

Measurement System Implementation

Before you begin to change the organization, ensure the project measurement system is in place. This can take place during or prior to the assessment. Early implementation of the measurements provides a benchmark of performance and sets the implementation team for a "win." I like to establish "scoreboards" for different areas and post the results on them. Each department has its own set of graphs. Sales might have graphs showing on-time delivery and orders booked. Manufacturing would have a scoreboard with the buffer management and constraint productivity graphs. A company scoreboard shows global measurements such as shipments, backlog, and lead time.

As you move through the early stages of the implementation, people are going to forget you are on a journey of greater than 90 days. Every time someone wonders, "Is this stuff working?" you can point to the measurements. This is the most critical step toward creating lasting support for your project.

Summary

Robust systems that are successfully implemented must consider both process implications and political ramifications. Do not forget the real-world implications of business process engineering. I do not profess to be a guru on navigating the shoals of organizational politics. However, you cannot ignore the perceived gains and losses a re-engineering effort will cause. Someone's "sacred cow" is going to get slaughtered (or they will think it is going to be), and you have to prepare for it.

The ideas of managing the control point and the ultimate location of the constraint involves a distinction that is not widely understood. Be prepared for substantial discussion and debate on the relative merits of various options. Even though I have discussed this concept in this chapter, I think it is best to have this discussion during the assessment stage to guide the system design. It is not as threatening at that stage, because the discussions are more abstract.

Be sure you understand management's thinking surrounding this issue before you make specific recommendations. Without this consideration, the direction of the implementation could be in opposition to the strategic direction of the firm, preventing a sustainable implementation.

7 Achieving Control

chieving control is the first prerequisite to implementing a process of ongoing improvement. When this step is completed, you will have the plant under control. *Under control* means that you can, with great confidence, predict what orders will ship on a given date or time. Poor due date performance (fewer than 95%), poor constraint utilization, and low (fewer than eight) inventory turns demonstrate the lack of control in a plant.

In most implementations, there will initially be an internal constraint that must be identified and broken before the organization can move on to identify and break external constraints. Therefore, most begin with a drum–buffer–rope implementation. The drum–buffer–rope scheduling system will give good control over the order fulfillment process, focus activity, and provide a foundation to gather information needed for ongoing improvement efforts.

Buffer management, which I will discuss in this chapter, is essential to delivering information to management about current and potential capacity constraints. Drum–buffer–rope is required to implement buffer management, so the first step is to implement a DBR system.

This chapter is about the detailed, practical aspects of implementations. You will have examples of actual policies and procedures that *have* been implemented and *do* work. The examples are just that. I have picked from a variety of implementations to show the principles — not the exact procedures (although you will be able to copy some of them exactly) — that you must follow for your business. With that caveat, let us begin.

Implementation Plan Overview

The plan overview is the most extensive of the implementation tasks (Figure 7.1). You are re-engineering the order fulfillment process. Order fulfillment

Process Step	Deliverable	System Component
Post critical measurements by department	Critical Measurements	Monitor and feedback
Set the constraint(s)/drum(s), establish capacity targets for the drum	Buffer Policy	Exploitation
Buffer sizes and policy are established	Buffer Policy	Exploitation
Develop scheduling policies	Scheduling Policy	Exploitation
Develop scheduling processes	Scheduling Process	Exploitation
Develop material/order release policies	Material Order Release Policy	Subordination
Develop synchronous manufacturing policies	Synchronous Manufacturing Policy	Subordination
Develop synchronous manufacturing process	Resource Allocation Process	Subordination
Specify queue areas and buffer locations on shop floor	Organized material storage	Subordination
Generic education of key staff, schedulers, and supervision	Workshops and seminars	Exploitation and subordination
Shop floor training	Workshops and seminars	Exploitation and subordination
Communicate new production policies and processes to supervision	Meeting	Subordination
Department Q&A sessions held	Meeting	Subordination
Schedulers are trained in new scheduling methodology and processes	Scheduler Training presentation	Exploitation and subordination
Quick response system designed	Quick Response Procedure	Exploitation and subordination
Implement First-In, First-Out on shop floor	First-In, First-Out Policy	Subordination
"Before You Accept the Order" checklist developed and implemented	Accepting orders checklist	Subordination
Review inventory policy on standard products and implement appropriate changes (as needed)	Inventory Policy and Procedure	Subordination
Change process to open orders implemented	Change Order Process/Change Order Form	Exploitation and subordination
Standard lead-time policy created and implemented	Lead-Time Policy	Exploitation and subordination
Manpower planning policy	Manpower Planning Policy	Subordination
Manpower planning process	Manpower Planning Process	Subordination
Process for daily/weekly capacity review	Action Meeting Process	Future constraint — elevation
Buffer management implemented	Buffer Management Procedure Buffer Analysis	Subordination and elevation
In-depth TOC practitioner training for schedulers (if required)	Workshops and seminars	Future constraint — elevation
Master scheduling process implemented	Master Schedule review process	Exploitation and elevation

Figure 7.1   Implementation Plan

crosses multiple functions and requires process consistency to operate effectively. The plan begins by establishing system measurements of the organization in order to evaluate whether the system is operating correctly. Next comes the development of policies and procedures to ensure that the control point (constraint) is being scheduled correctly. Following that are the subordination policies to synchronize the various functions. Last are the planning tools to monitor the system for movement of the constraint when the business changes (which it will).

You are about to do the detailed work of designing your new system. Each of the steps in the implementation plan calls for a documented procedure or policy. The purpose for documentation is primarily to achieve consensus on how you are going to run the business. If you can clearly articulate your process and policy, you have a much better chance of getting a lasting implementation.

Post Critical Measurements by Department

The "Critical Measurements" shown in Figure 7.2 are the first steps of measuring the major components of the order fulfillment process. These measurements should include the project measurements, but there are additional items related to the order fulfillment process that must be included, too.

Measurement	Description	Reporting Person	Posting Location
On-time delivery — % orders	Quantity of orders shipped on or before due date ÷ quantity of order lines shipped	Master scheduler	Materials scoreboard
On-time delivery — % dollars	$ value of orders shipped on or before due date ÷ value of orders shipped	Master scheduler	Materials scoreboard
Cycle time	Receipt of order (date) to ship (date)	Master scheduler	Manufacturing scoreboard
Sales shipped	Total sales $ shipped	Controller	Manufacturing scoreboard, break room
Sales backlog dollars	Total sales $ of open sales orders	Controller	Distribution
Manufacturing productivity	Throughput dollars shipped ÷ by plant operating expense	Controller	Manufacturing scoreboard
Constraint utilization	Actual hours run ÷ hours available	Production manager	Manufacturing scoreboard
Constraint rework hours	Constraint hours spent on rework × hourly throughput rate	Production manager	Manufacturing scoreboard
Constraint productivity	Quantity produced per hour of activation or number of standard hours earned ÷ number of hours worked	Production manager	Manufacturing scoreboard

Figure 7.2   Critical Measurements

The introduction of these measures is the beginning of creating organizational alignment. The objective of each is to improve the bottom line and the return on investment of the business. In other words, if any measurement is improved *independently*, it will result in an overall improvement of the business. If they all improve, then you are going to see wonderful things happen to the financial health of the organization.

The list is not inclusive. For example, you may want to add measurements for inventory, quality, or service. The idea is to implement a small number of measurements that monitor the critical aspects of the business (and the implementation), then communicate these to the entire organization. Think of it as your first step into a different way of managing the business, using more facts and less intuition. By sharing this information, you are about to become more inclusive in the management of the business.

Be sure to designate a person to put the measurements together, the frequency of update, and the locations these measurements will be posted. It does not matter who does it, although some may say that you do not want to have the department being measured reporting on its own performance. I think there are far too many checks on the system that prevent dishonest or distorted reporting, as long as it is clear how the numbers are derived.

As indicated in Figure 7.2, there are "scoreboards" where these measurements are posted. The scoreboards are a central place to post the numbers. It could be bulletin boards, white boards, or electronic signs. The idea is to have a place where you can communicate. Establish scoreboards in multiple locations to ensure wide dissemination and in department-specific locations to give feedback to the people accountable for the results.

On-Time Delivery

On-time delivery is the measurement of order fulfillment process control (Figure 7.2.1). If your process is under control, you will have a performance level of more than 95% on-time deliveries to the promised ship date. If you do not know what your performance is, estimate it around 50%. It is measured as both a percentage of orders (or order line items) and a percentage of dollars delivered on or before the promise date. It is the primary indicator of your performance in the customer's eyes and an important dimension of service, quality, and reliability. All things being equal, the greater your reliability, the greater your customer service.

Delivery performance is also a global measure of excess capacity. When delivery performance is very high, you have more than enough capacity to meet the current demands of the market. The order fulfillment system is

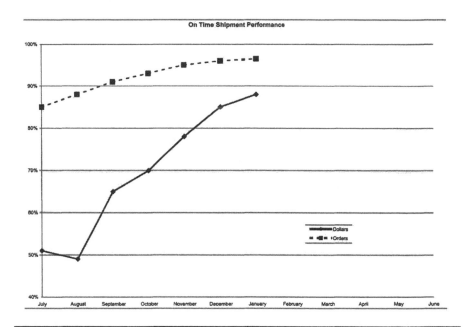

On Time Shipment Performance

Figure 7.2.1   On-Time Delivery

stable and under control. When the system is stable, the sales organization is free to stimulate demand to bring additional work into the plant. When delivery performance is declining or is poor, you do not have enough capacity to handle the current market demand. Sales should not stimulate additional demand. As *The Goal* demonstrates, plants often have extra capacity, but it is hidden through inefficient policies and practices; i.e., the constraint resources are wasted.

Both dollars and orders are measured to identify a common policy constraint that can have a dramatic effect on future throughput — the policy of pushing the high dollar value orders ahead of low dollar value orders. The order fulfillment process should work for *all* orders not just some of them. Small customers have a way of becoming large customers. If you cannot take care of them when they are small customers, what incentive do they have to become larger ones? A divergence in the percentage of dollars on time and orders on time is an indication that your priority system is not functioning properly.

Cycle Time

Cycle time (Figure 7.2.2) is a measurement of another critical service quality factor — responsiveness. This measurement represents the life of an order

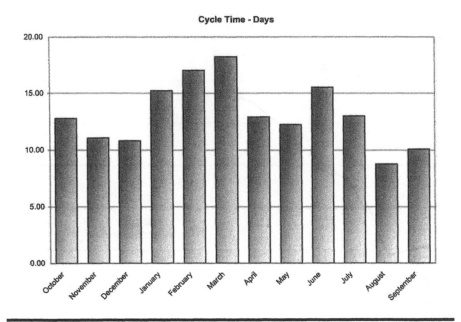

Figure 7.2.2 Cycle Time

from order entry to order shipment. Use it to determine the impact of your efforts on the customer. It is also an indicator of capacity availability. The greater the amount of protective capacity at nonconstraints, the shorter the cycle times.

Sales Shipped

The measurement of shipments to date (Figure 7.2.3) sets the expectation for everyone that it is a priority to deliver product to customers. I have worked in many organizations where this is not a given. Often, this group is reported to a small number of people or not reported at all (until the end of the period). Using the budgeted number establishes a goal for the organization to achieve. Remember that keeping score and telling people what the score is are two separate things. If you want to encourage appropriate behaviors to deliver revenue, you have to tell people how they are doing. That is not to say that everyone will make the connection between shipments and their daily activity, but this is the first step in your new game plan — posting the score.

Monitoring actual shipments vs. the goal will help to direct actions to smooth out the shipment flow and eliminate the "hockey stick" effect of shipping all of your revenues in the last week of the month. Use this measurement to evaluate how well the constraint is being used and that subsequent

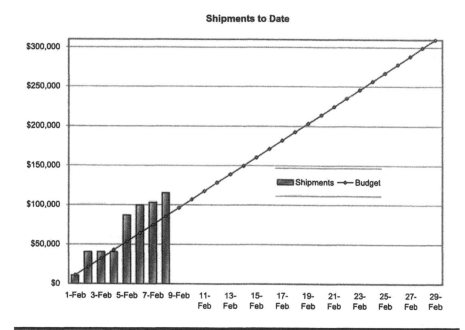

Figure 7.2.3   Monthly Shipments

resources are subordinated to it. If the product is not leaving as expected, it is an indication that your subordination processes are not working.

Sales Backlog

Sales backlog (Figure 7.2.4) is used primarily as an early warning that the plant is becoming overloaded or underloaded. Use it to identify the future state of the plant load. As sales backlog declines, become more aggressive in your sales efforts. If it is increasing, add capacity or find out why the plant is not keeping up with the market demand.

A secondary use of the backlog is to assess your efforts to break the internal constraint. Depending on from where you start, a declining sales backlog (assuming a constant rate of incoming orders) is an indicator that you are being successful in breaking the constraint. No change or an increasing sales backlog indicates your effort is not bearing fruit.

Manufacturing Productivity

Manufacturing productivity (Figure 7.2.5) measures how effectively the manufacturing organization produces results using the resources it has at its

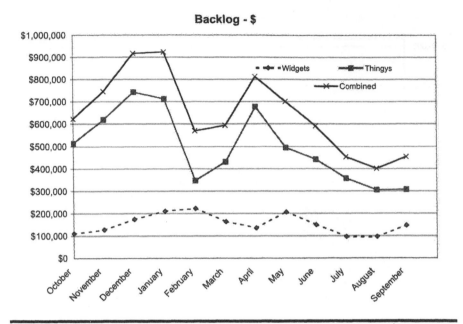

Figure 7.2.4 Sales Backlog Dollars

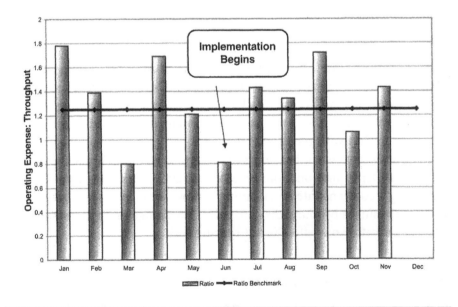

Figure 7.2.5 Manufacturing Productivity

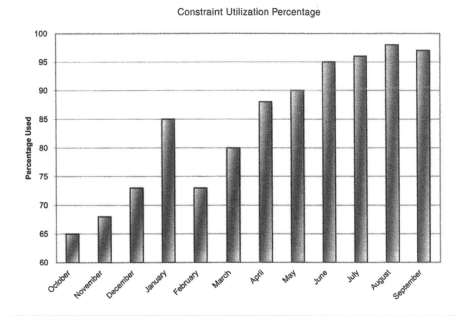

Figure 7.2.6　Constraint Utilization

disposal. Productivity is measured as a ratio of dollars expended to dollars shipped, i.e., operating expense to throughput. Therefore, if manufacturing becomes more effective (as defined above), it will produce more revenue for less expense, resulting in additional net profit.

Constraint Utilization

Constraint utilization (Figure 7.2.6) is measured to quantify the potential throughput lost by idle time. The constraint is being used only when it is producing product. This is a local measurement with global ramifications. Because the constraint is determining the rate at which you generate money, additional utilization will generate additional revenue (assuming it produces product for sale, not inventory).

Constraint Rework Hours

Constraint rework (Figure 7.2.7) represents lost throughput. Therefore, hours spent reworking are multiplied times the dollars of throughput produced per hour. In this way, the impact of lost constraint hours can be accurately reported.

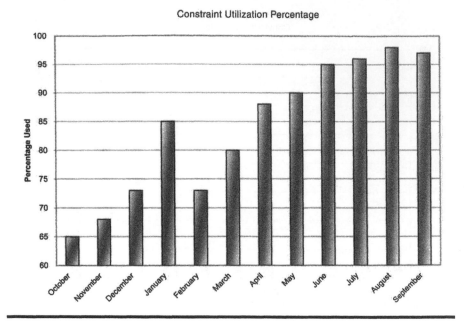

Figure 7.2.7   Constraint Rework

Constraint Productivity

Constraint productivity (Figure 7.2.8) measures how "efficient" the constraint resource is at producing according to a standard. I am not worried about earning hours (if there is a standard cost system), this measure is for trend analysis to see if you are becoming more productive. The three measurements of the constraint measure the different dimensions of resource performance: utilization, to see if you are using all the hours in the day; rework hours, to see if you are wasting them; and productivity, to see if you are getting better.

Set the Drum(s) and Buffers

The buffer policy document is your communication document for the location of the control points (drums), buffer sizes, and any special exceptions. In Figure 7.3, there are two different buffer sizes specified, depending on whether the product is a *standard* or a *new* product. This is to reflect the additional processing time in engineering for a new product. In the case of an engineer-to-order product, consider engineering (or design, art, etc.) as part of the production process. When the order is taken, you do not have all the detailed design work done. By the time you do, you cannot refuse the

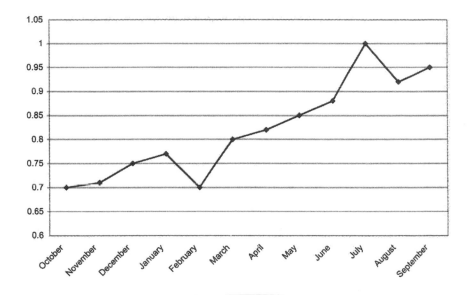

Figure 7.2.8   Constraint Productivity

Purpose
Communicate the buffer sizes in the various areas.

Definitions
A buffer is time to protect a resource (the buffer origin) from variation of the preceding resources. It is calculated by adding the process time for one batch, a time factor for "Murphy", and a time factor for resource contention.

Policy

	Standard Products	*New/Changed Products*
Drum Products		
A — Constraint	2 days	5 days
B — Assembly	2 days	5 days
C — Shipping	2 days	2 days
Non Drum		
B — Assembly	2 days	5 days
C — Shipping	2 days	6 days

Approvals
The scheduling manager, production manager, and the general manager must approve changes to this policy.

Continued

Figure 7.3   Buffer Policy

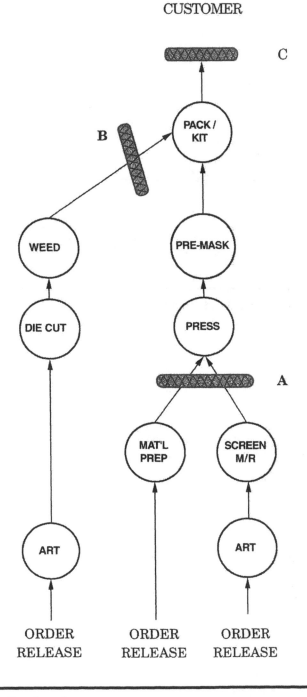

Figure 7.3 *Continued*

order and it is very difficult to change the terms. Therefore, the engineering is treated as a production function with specific times to develop manufacturing specifications.

There is also a buffer policy established for product that does not cross the drum resource(s). In the example, if products cross the assembly buffer, they are given the same buffer times. The shipping buffer specifies the minimum lead time.

This management policy should not be arbitrarily changed because it has such a big impact on the throughput of the business. That is why there are three approvals needed to change it.

Scheduling Policies

The scheduling policy document (Figure 7.4) is written to communicate the scope of authority and responsibility of the scheduling department. I first wrote this policy because the role of the scheduler is often misunderstood or poorly defined. The scheduler is your focal point for balancing what the customer wants and what resources you can bring to bear to satisfy those wants. Without a scheduling process, you will overcommit your resources, and poor customer service will be the result. You will not be able to manage your priorities very well. The best expeditors will get their orders out, while those who are remote from the plant will not. Either way, overall customer service will suffer.

Without a scheduling process, you will not have a plan to exploit the system's constraint. Someone has to do the detailed analysis to plan the work and then manage the plan. Without that, you will underutilize the constraint and you will lose throughput.

The role of the scheduler as the one person to commit the valuable resource of time on the constraint is akin to the role of the CFO in committing the financial resources. Poorly managed, this function can cost the organization in multiple ways. By closely managing the constraint, the scheduler can maximize revenue and customer service.

The policy lays out the reporting structure. In Figure 7.4, scheduling reports to the materials manager. I like to have the scheduler independent of manufacturing, to protect him/her from a bias to maximize resource efficiency. However, I have seen the scheduler report to manufacturing and be very effective. To whom the function reports is not as important as *defining* the function and having consensus that it is the correct structure.

The scheduling function is specified as the one to make realistic plans. A realistic plan is one that can be achieved. If production is consistently falling

Purpose
To ensure quick delivery of product without sacrificing promised shipping dates.

Reporting Structure
The scheduling function reports to the Materials Manager.

Functional Responsibilities
Scheduling has responsibility to:

Develop and communicate realistic manufacturing plans to the organization.

Resolve internal conflicts that prevent achieving commitments made to customers.

Ensure the plans to serve customers are executed.

Measure organizational performance to promised delivery dates.

Keep the organization informed of the status of customer orders.

Keep the organization informed of available capacity for sales.

Maintain an accurage database of open orders.

Keep the organization informed of potential shortages of capacity that could prevent a high rate of compliance to customer requirements and commitments.

Related Policies and Procedures
Buffer Policy

Manufacturing Synchronization Policy

Order Commitment Policy

Manufacturing Planning
Scheduling has the responsibility for managing the overall capacity plan to keep it in line with the demands of the market. In order to satisfy this requirement, scheduling reviews the sales and capacity plans weekly in the action meeting with representatives from manufacturing and sales.

Order Entry and Release
Customer promise dates and schedules ship dates are assigned by scheduling. This is done to maintain consistent delivery promises and provide visilbity to market or plant changes.

No orders are introduced into manufacturing unless material, tooling, and capacity are available (or will be available) to produce the order. No orders will be scheduled until all approvals and specifications are available. The schedule is frozen after the order is released into manufacturing.

Scheduling may, at its discretion, block out capacity for an order before formal order release.

Orders are not to be released early for the sole purpose of using manufacturing resources. People are to be moved to the work available at other resources or departments. Orders are released to manufacturing a buffer time ahead of need.

Scheduling
The scheduling function will maintain a realistic production schedule that ensures compliance with promised deliveries. Frequency of schedule update is done to help management understand the overall condition of the business (weekly minimum). Also, it should be flexible enough to accommodate changes in demands from customers and requirements from production. Once an order is released to manufacturing, the scheduled dates will not be changed. Those orders not yet released may be rescheduled as often as required.

The schedule(s) is/are the final authority on shop priorities.

Figure 7.4   Scheduling Policy

Execution
Manufacturing is accountable for conformance to schedule. Scheduling is to provide whatever support is needed to assist manufacturing.

Scheduling has the data to make recommendations to resolve conflicts between capacity and customer ship dates to ensure customer commitments are kept. They have the responsibility to recommend overtime and resource allocation in order to meet customer due dates.

In the event that scheduling cannot find an acceptable resolution to conflicts, the president or his appointee makes the final resolution.

Change Management
Scheduling is responsible for developing and administering effective procedures to manage changes to delivery dates, product specifications, or any other change that will affect open orders. This ensures changes are managed in an efficient, cost-effective manner that serves the customers. These procedures should include scrap replacement, rework, changes in product specification, changes in delivery requirements, and changes in order quantities.

Once and order has been introduced into manufacturing, the order will be completed. Requests for additional quantities or expedite orders must be put into production as soon as the gating operations can reasonably handle them.

Deviations
Changes to scheduling policy will occur with written authorization from the materials manager and the president. The materials manager can approve deviations to the policy (in writing).

Performance Measurements
The primary measurement is "on-time deliveries" (% of dollars and % of orders). Calculated and reported monthly as: Total dollar value of orders shipped on or before the promise (to the customer) date divided by total dollar value of orders shipped during the period. Also, total number of orders shipped during the period divided by the number of orders shipped on or before the promise (to the customer) date.

Figure 7.4 *Continued*

behind schedule, look at your scheduling processes. The scheduler is to take an active role in ensuring that the schedule is achieved. Schedulers become advocates for the production people when there are obstacles to schedule attainment. The scheduler is also the internal advocate for the customer. He/she is to work to remove the obstacles to delivering the product to the customer, whether it is a production or support department.

The scheduling department also communicates the status of customer service performance and available capacity for sales. The schedulers are the early warning system when the plant is starting to fill up or dry up. The scheduling department monitors the health of your order fulfillment system.

Scheduling policy is not set in isolation. It works with the buffer policies, manufacturing policies, and order commitment policies. These policies together set out your exploitation and subordination strategies.

The frequency of planning is spelled out. This will be a function of how volatile your demand is. Having said that, I do not think I have ever seen planning accomplished less frequently than weekly, although I have seen situations where planning meetings were held on a daily basis. In any case, if you do not have a formal system, spell out how often you are going to replan or make adjustments.

The policy also spells out how capacity is committed. Commitments are made in the form of promise dates to the customer. As orders arrive, capacity is committed. This is the same concept as "available to promise," except that instead of units, the commodity being promised is time on the constraint.

The release of manufacturing orders is also under the control of the schedulers. The timing and conditions are spelled out in the policy document. The guiding principles are to introduce into the plant only the orders that can be completed and only those needed to supply the buffers. Once started, the order is to be completed, even if you discover that the order delivery date has been changed. This is to protect the integrity of the priority system in the plant, where you will be following a first-in, first-out priority at all nonconstraint workcenters. You will not introduce work into the plant to activate nonconstraint workcenters. The excess work in process will increase lead times and ruin shop priorities.

The roles of the main players — scheduling and manufacturing — are spelled out explicitly. Scheduling sets priorities and manufacturing executes those priorities. Remember that scheduling is a supporting organization. This department has the responsibility to develop realistic plans and work with manufacturing and other departments to achieve them. In case of a conflict (nah, never happens), the resolution process is spelled out. In the example, the president has the last word. It may not be the president in your company, but it will be someone with profit-and-loss responsibility for the business being planned.

Be prepared for changes. Establish who will coordinate and execute them.

The primary accountability and measurement of scheduling is on-time deliveries. You could add a manufacturing expense component as well to provide balance. I did not add this because I think the plant manager is strong enough to override the scheduler's decisions when they are too expensive. Therefore, it was not needed. I think this is a personal preference. You choose.

Scheduling Processes

If the scheduling policy is a summary of your decisions of how the scheduling process fits into your business, the scheduling process is your detailed instruction to carry out those decisions. This is a training document, used mostly at the beginning of the implementation to ensure you have thought the process through and can communicate it effectively.

Figure 7.5 is a process to implement a manual drum–buffer–rope schedule. It is built on the five focusing steps. If you have a computerized system; you can skip these details. That does not mean you do not have to write the procedure, it means your process will be different.

The structure of the procedure document covers its purpose, special term definitions, resources required, people involved (and their accountability), the process itself, reporting issues, change protocol, and an example of how the dates work together (for nontechnical types like me who need pictures).

The scheduling process starts with establishing a detailed schedule for the constraint (drum) resource. The schedule should be done in time buckets no larger than 1 day. Forward schedule from today; you cannot do anything yesterday. You say the plant is behind? They are always behind? Get over it. The schedule is a plan, not a club to beat people with when they do not do what they should. If you need to catch up, the plan should reflect it. Scheduling work past due creates excess component inventory and destroys any credibility the schedule would have. Resist the temptation to schedule work past due. Every time you make a new schedule, adjust for the fact that the original plan was not perfect and start over. Every week.

Set the schedule priorities based on customer due date. I can hear the questions now. What about critical ratio and slack time and other methods to set priorities? Don't they give better indications of relative priority? Yes, they do. Do you want to take the time to calculate the critical ratio for every order in front of the constraint? No? Set your buffers conservatively and that will take care of your long-lead-time items. The short processing time orders will finish early and have taken little longer than what they could have been done in, but you will have a reliable process that is fairly simple to plan. Avoid sophistication in a manual process. It is much more difficult to manage.

Schedule the work that is physically present at the constraint first, for at least two planned buffers (if you have that much). This is a risk management strategy to avoid starving the constraint and making delivery promises that you may not be able to meet. Since the primary priority tool is the customer due date, if you have a need for an order that is not present, that order will have to be processed before it arrives. Scheduling these orders at the constraint less than one buffer time from today risks starving the constraint due to disruptions at resources before the constraint and missing the delivery date. The result of this step is a "frozen" schedule of the orders in the plant.

After the initial step, take the rest of the demand, backward scheduling from the customer due date, and fill the capacity of the drum resource. You will have conflicts among orders. Build early, add overtime, or reduce buffer size. The last option is to miss a delivery date. Each of the options has costs

Purpose
Document the process to be followed by the scheduler to produce realistic manufacturing schedules.

Definitions
A realistic schedule is one that reflects the actual shop capacity and actual customer demand and can be realistically accomplished.

Resources
Drum schedule(s)

Open order report

Schedule status including customer due dates

Inventory of all jobs in the buffer locations (Drum(s), Assembly, and Shipping)

Open unscheduled orders and capacity requirements

Capacity plan for next planning period (established in weekly action meeting)

Definition of Team
Scheduler (process owner): Crunch the numbers and distribute the production plan

Manufacturing manager: Validate process information — times, alternatives, resource availability (if needed)

Inside sales manager: Resolve conflicts among sales orders (if needed)

Process
1. Forward schedule the jobs physically in front of the constraint (drum resource), using customer due dates to establish priorities.

2. Continue scheduling forward until at least two planned buffers are filled or all orders identified in step one are scheduled, using customer due dates to set the sequence.

3. Backward schedule the balance of open orders in promise data sequence (ship date minus shipping buffer to establish due date at the drum).

4. Move jobs to resolve any conflicts between available capacity and required ship date.

5. If moving jobs does not resolve conflicts, consider other options (e.g., overtime, outsourcing, off-loading, reducing buffer time, or moving out ship date) to resolve any conflicts.

6. Estimate new ship dates (completion at drum plus shipping buffer plus process time after the drum).

7. In the event the new scheduled ship date exceeds the customer request date or promise date, notify sales immediately.

8. Schedule the release of material (start date at drum minus resource buffer minus process time for feeding operations)

9. Modify scheduled ship dates in system.

10. Release drum schedules to the organization.

Reporting
Report any orders that will ship late using a Late Order Notification form to sales and a summary to the management staff.

Changes
Requests for schedule modifications should be sent to the scheduler using a Change Request form.

The materials, manufacturing, and sales managers will make final approval for the schedule.

Figure 7.5   Scheduling Process

Buffers

Calculating completion dates:

Using the buffer sizes established in the buffer policy, subtract the buffer from the ship date, drum start date, or assembly start date. This will give the release dates. In cases when an order's process time at nondrum resources exceeds 8+ hours, an additional day should be added to the appropriate buffer.

This is an example of how to offset drum and release dates using buffers and a predrum process time that exceeds the 4-hour threshold.

This shows maximum protection against variation, assuming buffers are sized correctly. In case a customer requires a shorter lead time, the scheduler may reduce buffer time up to 50%. However, reducing the buffer time increases the risk of missing the promised ship date.

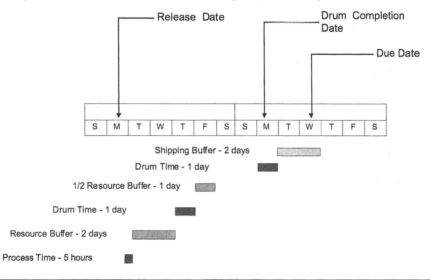

Figure 7.5 *Continued*

and risks. Increasing inventory or operating expense are obvious. When you reduce the buffer size, you increase the likelihood that you will miss a delivery date. Missing the delivery date has an immediate and long-term effect on throughput. Immediately, because you will not generate the revenue as quickly as you would like. Long term because if you cannot meet the delivery needs, customers will take their business to vendors that are more reliable.

After you have placed the orders on the timeline, add a buffer time to the completion date on the drum to arrive at a new ship date. Communicate changed ship dates to customer service or sales. Later orders have to be handled as a special case, and the customer must be contacted.

The diagram in Figure 7.5 shows the dates and the offsetting work to generate ship and release dates. You should have this diagram in your process, too.

Purpose

Ensure that material/order releases are performed in a manner consistent with requirements of the delivery and production schedules.

Policy

No orders are introduced into manufacturing unless material, specifications, tooling, and capacity are available (or are in line to be available) to produce the order.

Once an order is released into manufacturing it must be completed as planned, though scheduling may walk orders onto the schedule and/or drums if capacity and materials are available to meet rush order commitments.

Orders and material are released to manufacturing one buffer time ahead of need. Need is defined as the date at buffer origin (e.g., warehouse due date, assembly schedule date, or scheduled ship date).

Orders and material are not to be released sooner than one buffer time ahead of need just to increase utilization of manufacturing resources.

Deviations

Changes to this policy may occur with written authorization from the materials manager and the president. The materials manager can approve deviations to the policy (in writing).

Figure 7.6   Material Release Policy

Material/Order Release Policies

The material/order release policies (Figure 7.6) govern how and when work is released into the plant. This policy is the tool to get people to stop releasing work into the plant whenever they feel like it, thus distorting the priority system. The release policy is as important as your scheduling policy, especially if you have many nonconstraint resources.

The tendency in manufacturing is to work on the job that is the easiest, rather than the most important. Traditionally, we have used expediters to manage the queues. This is wasted effort. If you introduce the product into the system synchronized to the drum and implement the first-in, first-out policy, the amount of choice at the workcenter is limited to the jobs needed at the constraint or to support the constraint. When too much work is introduced, too much choice is given. It is then difficult to determine which jobs to work on first. That is when the priority system breaks down.

Synchronous Manufacturing Policies

The manufacturing policies (Figure 7.7) perform the same function as scheduling policies, to establish the role and mission of the manufacturing department. Their role should not change, but how they fit into the constraint management system has to be defined and agreed upon.

<u>Purpose</u>
This policy defines the mission of manufacturing.

<u>Accountabilities</u>
The primary accountability of manufacturing is to ensure that the department maximizes throughput dollars and manufacturing productivity. To this end, manufacturing is accountable to:

1. Maintain protective capacity at all resources.

2. Develop effective manufacturing processes that lead to greater manufacturing productivity (Throughput $/Operating expense).

3. Ensure that products meet or exceed customer expectations for quality, conformance to specifications, and delivery.

4. Maintain a trained and flexible workforce.

5. Maximize the use of materials, minimizing scrap and rework.

6. Keep the organization informed of the status of work in progress at the constraint resource(s) and buffers.

7. Keep the organization informed of available capacity and capacity limitations.

8. Inform scheduling daily of potential shortages of capacity that could prevent compliance with customer demand.

9. Ensure conformance to the first-in, first-out policy to manage the queues.

10. Post the current drum schedules at the constraint resources.

11. Move the people to the work.

<u>Manufacturing and Scheduling</u>
Manufacturing is accountable for communicating the overall capacity of the plant to scheduling. In order to meet this accountability, manufacturing will meet weekly with sales, engineering, and scheduling in the action meeting to ensure a reasonable balance between customer demand and capacity. New orders causing major shortages of capacity are dealt with in the action meeting.

Manufacturing is accountable for managing daily capacity to complete orders as promised. After release into manufacturing, interruptions to the plan may only occur at the direction of the manufacturing manager and the scheduling manager.

Manufacturing management may redirect orders to nonscheduled resources due to equipment breakdown or other reasons in order to minimize lost production time and maintain a high level of due date and drum date performance. Immediately upon such a change, scheduling must be notified in order to make appropriate adjustments to the schedules and notify sales of any resulting late deliveries (if necessary).

Manufacturing will monitor daily production through buffer management heetings, analyze planned versus actual buffer contents, and take appropriate action to ensure no starvation or waiting occurs at the drum resources.

Team members should be trained as backup operators for all resources.

<u>Analysis</u>
The manufacturing manager will meet with the manufacturing team on a routine basis (minimum monthly) to review manufacturing performance measurements, discuss quality issues, and address opportunities to improve constraint performance, reduce response times, and improve due date performance.

<u>Measurements</u>
1. Global measurements (monthly)

   a) Throughput dollars shipped Continued

Figure 7.7   Synchronous Manufacturing Policy

  b) Manufacturing productivity (T$/manufacturing operating expense)

  c) Manufacturing effectiveness (T$/constraint hours)

2. Individual group measurements (weekly)

  a) Drum efficiency

  b) Drum utilization (constraint hours spent on rework each week)

  c) Drum rework (constraint rework hours/week)

Figure 7.7 *Continued*

In most organizations, this will be obvious, but I have been in many where it is not. My views about the role of manufacturing are stated in the policy. Yours may be different. As long as everyone agrees, that is the main thing.

Synchronous Manufacturing Process

The synchronous manufacturing process (Figure 7.7.1) being described is not how to make the product, but how to allocate resources. The purpose is to establish the procedure to assess available and required capacities at nonconstraints, thus efficiently using production personnel. You do not want a nonconstraint resource to suddenly become or block the constraint.

Specifying queue areas and buffer locations on the shop floor sets out a visual monitoring system of the work-in-process areas. The idea is that you can informally manage the shop by managing the queues of work. If you see work accumulating somewhere besides in front of the constraint, it alerts you to a potential problem. Clear identification of where work belongs and where it does not simplifies managing the flow.

Conduct generic education of key staff, schedulers, and supervision. When I say generic, I mean "CM in production" education. Make sure the fundamental concepts are understood by the decision makers in the business, including front-line management and the supporting staff.

Shop floor training is a necessity. Making the transition from traditional practices to drum–buffer–rope scheduling will mean a dramatic drop in work-in-process inventory. Most shop floor people associate high work-in-process inventories with job security. When the shop dries up, fear will be created and productivity will suffer. In order to prevent this effect, we educate people on the mechanics of drum–buffer–rope. If they know what is coming, they are less likely to resist the change or overreact by slowing the pace of work.

Purpose
Assess available and required capacities at nonconstraints, thus efficiently using production resources.

Resources and Information
Buffer management reports — which nonconstraint resources are causing buffer shortages.

Queues at nonconstraints — visual observations of resources lacking capacity.

Schedules and dispatch lists — identification of what is coming.

Accountabilities
The manufacturing supervisors are accountable for the procedures and their implementation.

The manufacturing supervisors will make formal reviews of in-plant queues weekly or on an as-needed basis.

Procedure
1. Scheduling will release work orders to manufacturing a buffer time ahead of the need date, according to the material release schedule.

2. Manufacturing will process all orders at nonconstraints first-in, first-out, and monitor queues on the plant floor.

3. At the beginning of each shift, manufacturing supervisors will meet to review queue status, identify buffer penetration conditions, and extra capacity and to prepare action plans to resolve conflicts between available capacity and completion schedules. (Immediately following the buffer meeting.)

4. When the queues in the plant or at specific resources decrease, and/or the anticipated demand decreases, those operators are moved to other resource areas that are experiencing increased loads. Early release of work is a *last resort* and is subject to approval of the scheduling team.

5. When the queues in the plant and/or the anticipated demand increase, capacity is added from other less loaded resources, outsourcing, overtime, etc.

6. When queues are consistently high over a period of 6 weeks, additional permanent capacity is to be considered.

Figure 7.7.1   Synchronous Manufacturing Process

Communicate new production policies and processes to supervision. Ensure the front-line leadership understands what the system requires. Remember they used to be focused on managing resource efficiency and utilization. Now they are to be focused on delivering product to the buffer on a timely basis, i.e., managing the flow.

In addition to the conceptual training, department question-and-answer sessions are held to explain the new measurements, procedures, and practices of the new system. This is to avoid problems and give a maximum chance of successful implementation.

Schedulers need to be trained in new scheduling methodology and processes. The DBR scheduling methodology has to be taught. Your design and the size of the organization will dictate what this training will be. It could be as simple as spending a few hours with the scheduler, or the training could

Purpose

Establish a system by which orders that are needed in less than the scheduled lead time may be evaluated, scheduled, and produced.

Resources and Information

■ The standard lead-time report

■ Complete product specifications

■ Bill of material for job

■ Customer required date

Procedure

1. Salespeople requiring ship dates less than the reported standard lead time will forward the completed order specification sheet along with the requested delivery date to the inside sales manager for evaluation and approval.

2. Once approved by the inside sales manager, the scheduler will review the order to evaluate capacity requirements and availability then create a plan to achieve the needed delivery date.

3. Assuming capacity is available to produce the order, the scheduler will then work with materials staff to identify any special parts requirements or parts shortages.

4. The result of the analysis will be an evaluation of the impact of shipping the product in less than the current lead time.

5. In the event the order requires additional operating expense or increased material cost to process, the order will be reviewed with the inside sales manager prior to acceptance.

6. The inside sales manager will then review the pricing and determine if the quoted price is acceptable.

7. The inside sales manager will promise the delivery date to the salesman.

8. The order follows the normal order process with the required delivery.

Figure 7.8   Quick Response Process

be a seminar given to a group. I like to cover all the policies and procedures for each of the areas, as well as the mechanics of how things are to be done.

Other Processes and Policies

Quick Response System

No matter how robust your system is, there will always be customers who demand delivery in less than the current lead time (Figure 7.8). Recognize it and develop a procedure to shortcut the system. Sometimes, it helps to break the products into categories to simplify management of the system.

First-In, First-Out on Shop Floor

An important part of subordination is the implementation of first-in, first-out at the nonconstraint resources (Figure 7.9). In the case of a shop that

<u>Purpose</u>
Maintain consistent job priorities at all resources in the plant and compliance to promised ship dates.

<u>Policy</u>
Jobs are to be worked on first come (first-in), first worked (first-out) basis at all nondrum resources.

<u>Exceptions</u>
Jobs to be reworked, whether in house or customer returns, are to be worked next up.

Jobs late or in danger of not making the scheduled ship date or drum date are to be moved to the next job up and worked in sequence. These jobs will be identified either by the scheduler or in the buffer management meeting.

<u>Accountabilities</u>
Scheduling is accountable for the release of jobs into manufacturing.

Manufacturing is accountable to ensure that all work at all resources is processed on a first-in, first-out basis.

Figure 7.9   First-In, First-Out Policy

has many jobs arriving in a day, a method has to be devised to determine the sequence in which the jobs arrive. This will dictate the sequence of processing. The procedure addresses two things — how to know what arrived first, and how to make sure people work in that sequence.

"Before You Accept the Order" Checklist

In make-to-order environments, the proper discipline to completely specify the order before production is sometimes lacking. To promote the shortest cycle time possible, the product and order must be completely specified before being launched into production. This policy spells out what needs to be defined before you declare an order as accepted (Figure 7.10).

The order commitment policy (Figure 7.11) is a supplemental document that spells out the date commitment policy. Sometimes, there is significant disagreement about what promises are to be made to the customer. This document spells out the agreement on what and how capacity is promised.

Review Inventory Policy

Depending on your buffering strategy, the stock levels and ordering policies of standard products may be modified. Your implementation may call for stocking certain kinds of products to promote short lead times or stocking product at an intermediate-manufacturing step. The objective of the policy

Purpose

To promote the shortest cycle time possible in manufacturing by completely specifying customer requirements before order entry.

Definitions

An order accepted is one in which any department must spend time and/or money to fulfill the customer's expectations.

Accountabilities

The salesperson is responsible for gathering the required information from the customer and coordinating activity within the departments to ensure the order is "captured."

The sales manager is responsible for price approvals.

Application engineering is responsible for creating the detailed product specifications for manufacturing and deriving the product cost information.

The controller is resonsible for assessing the credit status of all new customers and for the creation of payment terms.

Scheduling is responsible for delivering accurage lead-time information to the salesperson/sales department.

Process

The following questions must be answered affirmatively on each order by the sales staff before an order is accepted. If any answer is negative, the salesperson must obtain the appropriate department head approval.

- Are the product and order fully specified? Understand the specifications. Refer to the specification checklists to identify all needed information. Any open issues should be referred back to the customer before order entry.

- Do our current lead times meet those needs? Can we do everything required to make the product in the length of time the customer will accept? If the requested lead time is too short, thereby requiring overtime, will the customer pay a 35% up-charge?

- After design/price quote submission to the customer, have we met the customer's budget requirements? Will the customer pay our quoted price?

- Is our quoted price within our cost of goods/margin/throughput guidelines established for each operating area? Can we make what the customer wants at the price we require?

- Is the customer creditworthy? The controller must approve special terms and new customers before order entry.

If all answers are yes, then we may accept the order. If any answer is no, then the salesperson must work with both customer and department heads to arrive at a satisfactory compromise.

Figure 7.10   Before You May Accept the Order

is to ensure that the inventory levels are reviewed and monitored on a regular basis to maintain consistency-with-demand patterns and allow management to make informed decisions about inventory investment (Figure 7.12).

The inventory procedure (Figure 7.13) spells out who does what and how often regarding the management of inventories. In a larger organization, this will probably be split among several people.

Purpose
To promote a level loading on the manufacturing, materials, and engineering resources.

Measure performance against customer's expectations.

Ensure accurate delivery commitments are made at order entry.

Definitions
Customer request date — The date the customer indicates as the "need" date for the order.

Promise date — The date promised or committed to by us to our customer. This is the date used to calculate on-time delivery/due date performance.

Scheduled ship date — The date assigned by scheduling from the current manufacturing schedule.

Accountabilities
Since it is our goal to exceed the expectations of our customer, we will always get from the customer a "need date" or "delivery expectation." It is a primary accountability of sales to have a dialog with the customer to establish the need date, rather than pull dates from the lead-time report. On-time delivery performance is as much a factor of the initial setting of promise dates as it is of any of the processes involved in completing the order.

Sales will keep our customers advised of any changes in our scheduled ship dates that adversely affect the promise date.

Scheduling will furnish guidelines regarding current manufacturing schedules, lead times, and available promise dates to the sales and sales support departments on a daily basis.

Every attempt will be made to esceed our customer's needs and expectations.

Figure 7.11   Order Commitment Policy

Purpose
Minimize inventory while supporting internal and external product demand.

Definitions
Inventory — The amount of money invested in things intended for resale. The TOC view of inventory includes materials, parts, machinery and equipment, tools, supplies, and even facilities as inventory, as they ultimately are intended for resale. For this policy, inventory is defined as purchased raw materials, work in process, and finished goods.

Stock inventory items — Materials, parts, finished goods, and supplies maintained for support of manufacturing, sales, engineering, and/or office support, which are set up as stocked items with part numbers, reorder points, reorder quantities, and assigned inventory locations and are subject to automatic reorder and replenishment in the inventory control system.

Nonstock inventory items — Materials, parts, finished goods, and supplies which may on hand, but not subject to automatic reorder or replenishment.

Inventory superintendent — The person assigned daily responsibility for the physical management of an inventory item group.

Inventory specialist — The person assigned the daily responsibility for managing the ordering and overall levels of inventory.   Continued

Figure 7.12   Inventory Policy

Policy

Inventory levels are to be held to a minimum amount such that they provide sufficient support for manufacturing, sales, engineering, and office functions without creating shortages for internal or external market demands. Shortages are defined as:

◼ Parts missing for shipments relative to promise dates

◼ Parts missing from the constraint and shipping buffers

◼ Parts missing that support sales, engineering, and office functions

Inventory is considered a liability rather than an asset, due to not only the cash it ties up, but also the carrying cost, the likelihood of becoming obsolete, shrinkage, and opportunity cost. While we recognize that the relationship of inventory investment to net profit, cash flow, and return on investment is generally negative, inventory levels must be maintained high enough to prevent limiting the organization's throughput.

Inventories are to be adjusted as levels of sales and product mix change to provide reliable support for the current market mix.

Obsolete inventory is to be disposed of at the best price attainable as soon as it becomes obsolete.

Nonstock inventory items are ordered "as needed" determined by job requirements and/or supplier minimum order quantities.

The materials manager, purchasing manager, and the inventory specialist are responsible for the monitoring of late shipments, buffer contents, and other shortages causing interruptions in the workflow due to part shortages. The buffer management meeting, the weekly action meeting, line-item usage data, and the sales forecasts are tools for use in managing inventory.

Any department manager may submit requests for additional stock inventory items. The inventory specialist, subject to approval of the materials manager, sets up stock inventory items.

Review/Analysis of Inventory Levels

The inventory specialist is responsible for the management of stock inventory items, including:

1. Training of inventory superintendents,

2. Evaluation of inventory superintendents no less than once each quarter, and

3. Records management of stock inventory items as directed by the materials manager.

Stock items are reviewed weekly for reorder.

When appropriate, stock inventory items are stratified into importance categories of A, B, and C. Stratification allocates different levels of attention, with A being the most closely scrutinized and C being least scrutinized.

A items are those items for which out-of-stock positions create serious problems with on-time performance and/or customer relations or require significant investment.

The inventory specialist evaluates stock inventory items at least once per quarter. Items for review are order frequency and annual usage, with reorder point and reorder quantity adjustments being made as indicated by the reviews.

Measurements

Inventory turns are measured monthly.

Buffer shortages caused by inventory are measured in dollars assigned to inventory from the Buffer Management procedure.

Deviations

Changes to this policy will occur with written authorization from the materials manager and the president. The materials manager can approve temporary deviations to the policy (in writing).

Figure 7.12 *Continued*

Purpose
Minimize the investment in finished goods and component inventory while delivering superior service (product availability) to our customers.

Resources
Reorder report.

Procedure
1. Stock items are reordered by the inventory specialist on the 1st and 15th of each month.
2. Stock inventory items are stratified into categories of A, B, and C for reorder point (ROP) and reorder quantity (ROQ) adjustments

   A category items, to be analyzed semi-weekly, include those critical items for which out of stock positions create serious problems with on-time performance and/or customer order promise dates. A category includes those itmes that meet any of the following criteria:

   Lead times greater than 10 working days

   Minimum or practical order values exceeding $5,000

   Vendor performance that varies by more than ±5 working days

   B category items, to be analyzed monthly, include those items that meet either of the following criteria:

   Lead times greater than 5 but less than 10 working days

   Minimum or practical order values greater than $1,000 but less than $5,000

   C category items, to be analyzed quarterly, include those items that meet either of the following criteria:

   Lead times less than 5 working days

   Minimum or practical order values less than $1,000.
3. Stock inventory item levels are calculated using order and shipping time, pulus a minimum of two weeks usage. Actual levels will depend on dollar value, quantity pricing, and the stratification category.
4. All stock inventory items are analyzed at least quarterly for order frequency and annual turns, with ROP and ROQ adjustments being made as indicated by the analysis.

Accountabilities
The materials manager is accountable for the overall inventory policy and performance.

Figure 7.13   Inventory Procedures

Change Process to Open Orders

In-process orders will be changed. Make sure your system design allows for it (Figure 7.14). Who should be allowed to change an order? For what kinds of changes? Who should be notified? These are the kinds of issues to address.

Standard Lead-Time Policy

A procedure to promise delivery dates is as important to on-time delivery as good scheduling. This policy defines how to use the current manufacturing

Purpose
Ensure that changes to open customer orders are performed accurately while maintaining promised ship dates.

Definitions
Open order — An order becomes an open order once sales has completed and printed the job sheet and remains open until it is shipped and invoiced, or canceled.

Released order — A released order is an open order that has been released to manufacturing but not completed.

New order — Same as an open order.

Completed order — New order that has been shipped but not invoiced.

Finished order — Completed order that has been invoiced.

Resources
■ Change Order form

■ Open Order Report

■ Shipping Buffer Report — late orders

Process
Before order release, changes to open orders may be made verbally or via e-mail between scheduling, sales, and/or engineering. After the order has been released, all changes (with the exception of scheduled ship date changes on late orders) are to be documented using the Change Order form. These include changes to promise date, quantity changes, and any cancellations.

Promise Date Changes at Order Entry
1. Lead-time driven (before order release) — Upon receipt in scheduling, the promise date on the new order is reviewed to assure conformance to current manufacturing capacity and lead time as well as purchasing lead time of any nonstock items.

2. If the promise date is determined to be unattainable, then scheduling will change the promise date and schedule the order to the new date.

3. When scheduling changes the promise date in the system, the scheduled ship date will automatically change to match the new promise date.

4. Scheduling will immediately notify the salesperson responsible via e-mail or by completing and hand-delivering a change notice.

5. Sales will then contact the customer to confirm that the revised promise date will still satisfy its need.

6. If there are problems with the changed date, sales should request a meeting with scheduling and manufacturing to review the conflicts and look for alternatives (increase capacity — overtime, schedule change — move out other jobs).

7. Once a resolution is negotiated, scheduling will enter the appropriate promise date and scheduled ship date, and reschedule the order according to that date. Sales will confirm with the customer.

Customer Driven
1. Using the Change Order form sales will report to scheduling any date changes to open orders.

2. After notification, scheduling will review the order status. In instances where the promise date is being moved up, scheduling may need to confer with manufacturing to determine the feasibility of the request.

Figure 7.14 Change Order Procedure

3. Scheduling will then report to sales on the status of the Change Order. The change will be:

   a) Accepted with no change, or

   b) Modified (date change made other than request), or

   c) Denied (date change is not feasible). Any rejections are subject to the appeal process described earlier.

Scheduled Ship Date Changes

1. The manufacturing, scheduling, and engineering departments, during the daily buffer meeting, review the order status.

2. Late orders and potential late orders are identified, problems resolved, and corrections made.

3. Scheduling will respond to sales with a daily late order report notifying sales of a late order, the reason for the late order, and the revised scheduled ship date.

4. Any scheduled ship date changes requested by the customer and accepted by the company will be treated as promise date changes.

Quantity Changes, Cancellations

Sales will report to scheduling via the Change Order form any quantity changes or cancellations requested by the customer.

After notification, scheduling will review the order status. In instances where work has already begun and material orders have already been placed, scheduling will follow up with purchasing and manufacturing on the feasibility of the quantity change/cancellation.

Scheduling will then report to sales on the status of the change order including any charges incurred if canceled.

Upon agreement and acceptance of a change order by all (sales, manufacturing, scheduling), it is the scheduler's responsibility to locate the job ticket and all pertinent job splits to make the changes to this paperwork along with any changes to the schedule which are needed. Scheduling is also responsible for inputting date changes to the computer.

The sales person is responsible for updating the computer of any cancellations or quantity changes.

Once processed by scheduling, completed change orders will be attached to the job sheet.

Figure 7.14 *Continued*

lead times by sales and customer service to inform customers of current delivery lead times (Figure 7.15). By promising deliveries based on current backlog, you enhance your ability to keep your commitment and increase customer satisfaction. The backlog of work is calculated at the drum resource. If you have multiple drums, you will have multiple lead times affecting your promise dates.

Manpower Planning Policy

In many businesses, capacity is determined not by machine availability, but by people availability. The objective of this policy is to ensure plant staffing is sufficient to satisfy market expectations (lead times) and maintain maxi-

Purpose
Define the use of standard manufacturing lead times by salespeople as a general guide to inform customers of current manufacturing lead times. Promote accurage quotation of order delivery dates.

Definitions
Standard manufacturing lead time — The total time from the receipt of the order in the scheduling department ot the ship date. This period does NOT include specification, drafting, customer approval time, or procurement time for nonstock items.

Policy
Calculating Lead Time
Manufacturing lead time is calculated using the scheduled backlog then adding 1 week for in-process jobs not yet scheduled. For example, if the backlog of work is 8 weeks, the lead time will be quoted as 9 weeks.

Quoting Lead Time
Lead time will be quoted at least weekly via e-mail by the scheduler. All salespeople and the management team will be copied. Lead time will be reported separately by product category. Additional categories will be added as necessary.

Order Promising
Salespeople will use the lead time report to promise deliveries to the customer. In ALL cases, the salesperson should commit orders based on the customer's need date. In the event the customer does not know when he needs the order, the standard lead time should be promised. In the event the lead time is greater than the customer's need date, the Quick Response Procedure should be employed. No order can be promised for less than the promised lead time unless the inside sales manager gives written authorization.

Database Maintenance
The scheduler will be responsible for maintaining accurate due dates in the Notes and ERP database. These dates MUST match: Pack date (Lotus Notes), the Customer Order (C.O.) completion date, and the Work Order (W.O.) completion date (in ERP). The dates will be updated every time the production schedule changes.

Accountabilities
The scheduler is accountable for the creation and distribution of the lead-time report with revisions as the situation dictates.

The inside sales manager is responsible for initializing the quick response procedure.

The salesperson is accountable for the use of the reports when negotiating customer request dates and promise dates on customer orders.

Figure 7.15   Standard Lead-Time Policy

mum return on investment of plant equipment. The planning policy sets forth the guidelines for staffing (Figure 7.16).

Manpower Planning Process

This spells out the procedure to determine when to add and reduce labor capacity in line with the policy (Figure 7.17). The idea is to hire only when

<u>Purpose</u>
Ensure plant staffing is sufficient to satisfy market expectations (lead times) and maintain maximum return on investment of plant equipment.

<u>Policy</u>
Staff and manage the plant so that the boring mill resources operate as the manufacturing control point resources.

Provide sufficient protective capacity at all other resources to absorb fluctuations in market demands (mix changes), with recovery time from peak loads being held to no more than 2 days.

Maintain a manufacturing productivity ratio of 9.35.

Cross-train sufficient personnel to provide immediate additional capacity for resources that have the least protective capacity during peak loads.

Manage the daily assignment of personnel to ensure that, at the beginning of each shift, there is no more than 1 day's work in front of all nonconstraint resources, and/or that the recovery time from peak loads is minimal (1 to 2 days).

Continually train supervisors and existing and new employees in the principles of the constraint-based philosophy of the management of our business.

Identify and create career paths for high performers.

Anticipate the effects of future growth and develop specific contingency manpower plans to accommodate likely future market demand shifts or key orders, as identified by sales.

<u>Accountability</u>
The manufacturing manager is accountable for executing the policy.

<u>Measurement</u>
The effectiveness of labor planning will be the manufacturing productivity ratio. This ratio is the relationship of dollars spent in production to the throughput (sales – total variable cost) dollars generated (shipped) during the period.

Figure 7.16   Manpower Planning Policy

recovery from peak loads at nonconstraints exceeds some specified period and/or lead times exceed market requirements. When lead times are shrinking, pressure is applied to sales. When they are increasing, pressure is applied to manufacturing.

Daily/Weekly Capacity Review

The daily or weekly capacity review is to keep the short-term sales and manufacturing plans synchronized (Figure 7.18). I call it the action meeting, because that is the result. If you do your job properly, there will be something happening each week to respond to changes in the market. The degree of change will be dictated by how well you do your long-term planning. Just make sure the meeting does not become an analysis meeting. Come to the meeting with the analysis, then make decisions based on what is presented at the meeting.

<u>Purpose</u>
1. Staff and manage the plant so the named drum resources operate as control point resources.

2. Provide protective capacity at all other resources sufficient to absorb fluctuations in market demand.

3. Maintain a manufacturing productivity ratio of 9.35.

4. Have personnel to support resources that have the least protective capacity during peak loads.

5. Manage the assignment of production personnel.

<u>Resources</u>
1. Existing manpower resource availability

2. Workforce skill definition — who can do what

3. Current order backlog — from the drum schedule and the open order scheduled ship date report

4. The buffer management report — identification of which resources are causing holes in the buffer

5. Queues — work-in-process inventory; daily visual observation of resources facing more work than can be completed in 1 day

6. Forecast of type and amount of future sales and/or market demand — from the weekly action meeting

<u>Process</u>
Recruit and hire reliable employee candidates with enthusiasm, good work ethics, and the ability to see the "vision," who can and will contribute to moving the company closer to the goal.

Continually train supervisors and existing and new employees in the principles of constraint management and the management philosophy of our business (interactive game training and daily use of the TOC language and teaching).

Actively participate in the weekly action meeting, collect and analyze the factors itemized above (1 to 6).

Move people resources to the machine resources that have the largest queues.

Anticipate the effects of future growth and develop specific manpower plans to accommodate likely future market demand shifts.

Focus on the five steps of constraint management.

Hire as necessary:

■ When recovery time consistently (over a 3 to 4 week period) exceeds the standard

■ When buffers are repeatedly penetrated by non-control-point resources

■ When it is anticipated that mix fluctuations will cause a capacity constraint

Figure 7.17 Manpower Planning Policy

Buffer Management

Buffer management is major milestone number 2. This procedure is as critical to the long-term success of the implementation as the schedule of the constraint is. Buffer management is the procedure to analyze and control the

Purpose
Keep the short-term sales and manufacturing plans synchronized.

Identify changes in market demand or plant capacity and create an action plan to respond.

Team
Scheduler or order fulfillment manager (process owner): Responsible for reconciling conflicts between sales forecast and production plans and managing the planning process to ensure the organizational objectives are met.

Sales Manager: Responsible for maintaining and communicating the sales forecast.

Manufacturing Manager: Responsible for managing capacity to achieve the forecast and for communicating changes in resource availability and capability.

Engineering Manager: Responsible for managing engineering resource to achieve the forecast and for communicating changes in resource availability and capability.

Resources (for the next 4 weeks)
Plant capacity vs. load information

Budgeted capacity vs. actual capacity

Current drum schedules — Constraint schedule to show backlog

Information on upcoming sales (forecast)

Open order reports

Unrouted jobs list

Process
Review status of action plan from last week.

Review status of current backlog, capacity status, and forecast sales activity.

Identify problem areas; i.e., conflicts between forecast and capacity.

Establish drum capacities for next 2 weeks (plan production levels).

Develop action plan and accountabilities.

Adjourn.

Publish minutes (decision and action plan).

Figure 7.18   Action Meeting Process

content of the buffers and correctly allocate resources. Done well, you will maximize throughput, minimize response time, and maximize on-time delivery performance.

Buffer management is, at first, a procedure to validate your buffer sizes. Remember, at the beginning of the implementation the sizes are estimated (conservatively). Therefore, you must go back and check to make sure they have been established correctly. The procedure is also a day-to-day management tool to resynchronize the schedule and stay on plan. The most lasting benefit is as a diagnostic tool to identify shortages of capacity and plant "hot spots" for improvement activities.

The Drum Schedule

Day	Part	Qty	Hrs
1	Z1	25	5
1	Q5	5	3
2	Q5	5	3
2	W3	5	5
3	W3	2	2
3	R9	2	6
4	R9	1	3
4	Z1	25	5
5	W3	22	2
5	Q5	10	6

Figure 7.19.1 Planned Buffer Content — Day One

The Time Buffer

While the plant operates, inventory accumulates in front of various resources as the tasks (jobs) wait for processing. The resources that consistently have the largest accumulation of inventory are likely candidates for the constraint(s).

Since the constraint is the resource that has the *least* capacity relative to the market demand, all other resources have *more* capacity relative to the constraint. Therefore, you should have a queue of tasks ready for processing by the constraint resource. The queue does not contain random jobs; it contains the scheduled jobs for the constraint (drum). If material is being released as a function of the drum (as it should be), the queue will contain the next scheduled tasks for the constraint.

This queue of the next scheduled tasks composes the *buffer* (Figure 7.19.1). Some of the buffer will (or should) reside at the resource ready for processing at all times. It is called the buffer because it "buffers" the constraint against fluctuations in the system. As the name implies, fluctuations at upstream resources will cause the queue in front of the constraint to fluctuate. As time passes, the constraint depletes the physical queue and the feeding resources replenish the buffer. It is a revolving stock of work for the constraint — the content is always changing (Figure 7.19.2). Since the components of the buffer represent available work for the constraint, we refer to it as a *time buffer*. The measurement of the buffer is *always* time.

This planned buffer content allows us to predict when jobs will arrive at the constraint. More importantly, it allows us to determine when tasks are *not* at the constraint. These missing tasks, or *holes* in the buffer, tell us when and to what degree there are fluctuations in the system feeding the buffer.

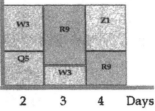

The Drum Schedule			
Day	Part	Qty	Hrs
1	Z1	25	5
1	Q5	5	3
2	Q5	5	3
2	W3	5	5
3	W3	2	2
3	R9	2	6
4	R9	1	3
4	Z1	25	5
5	W3	22	2
5	Q5	10	6

Figure 7.19.2   Planned Buffer Content — Day Two

Quantifying Variation

The buffer guards against the process variation preceding the buffer origin (in this case, the drum). Since variation is always present, we will never see the *entire* planned buffer physically present at the constraint resource. Ideally, we should find about half of the planned buffer as work ready to be processed (Figure 7.19.3). If there is more than half, the buffer is too large. The fluctuations in the system are not great enough to warrant the investment in inventory.

If there is less than half, the buffer is too small. The throughput of the system is in jeopardy. The risk of the constraint being idle increases as the physical portion of the buffer decreases. You should think of managing the buffer as managing risk — the risk of stopping the throughput of the organization.

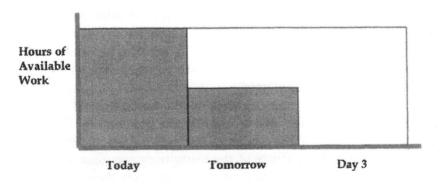

Figure 7.19.3   Ideal Buffer Content

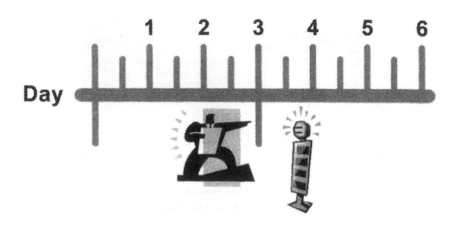

Figure 7.19.4   Red Line Indication

Predicting Problem Areas

When the buffer falls below the ideal, it is a signal that there is excessive variation in the system. When there is excess (greater than expected) variation, it is a trigger for action. Depending on the depth of the penetration of the hole into the buffer, you should check status, expedite, or reschedule. Do this before the constraint is idled. The buffer is an early-warning system for the entire system, warning of potential problems in output.

To track disruptions, I prefer the method described by Eli Schragenhiem to the method described by Eli Goldratt. Rather than dividing the buffer penetrations into three regions, it is divided into two. Figure 7.19.4 shows a 6-day buffer with a 3-day (half the planned buffer) look ahead. When a task is not present inside the 3-day look ahead, it triggers an action and a recorded disruption to the buffer (red line or light). I have found this method much more practical, because no one really looks at anything outside the first two regions anyway. So, if a task is missing inside half the buffer, action is taken. Outside half the buffer, who cares?

Usually, when there are penetrations into the buffer, the missing tasks will be found at the resource that is causing the penetration. Troublesome resources will frequently cause holes in the buffer. If we monitor and track the penetrations and the source of the disruptions over time, patterns will emerge. Some resources will show up more frequently than others as troublesome. We can then address those most troublesome nonconstraint resources, adding capacity or otherwise resolving the excessive variation problem, thereby reducing the fluctuations, which then allows reduction of the size of the buffer. Once the buffer is reduced, we can begin to monitor the fluctuations and disruptions again, then focus our attention on those resources that are creating the greatest risk to throughput, continuously improving the process.

Calculating the Buffer

The buffer is equal to process time of one part plus a time for normal variation and a time to accommodate resource contention. Process time should include transportation, actual process time, and setup time — nothing else. Normal variation time (Murphy's Law) is the time it takes to recover from a typical problem. Resource contention is an added time to account for the noninstant availability of resources when the task arrives.

These rules of thumb will allow you to set the buffer initially. The only effective means of determining the "correctness" of the buffer size is to actively manage it. Buffer management has two components: a daily meeting and a reporting process. The daily meeting determines the short-term action plan to achieve on time delivery. The reporting process shows trends toward resources becoming bottlenecks.

The buffer penetration graph (Figure 7.20.1a) or the buffer penetration table (Figure 7.20.1b) is the tool to communicate the resources that are the most troublesome, i.e., that cause the most "almost late" (those that have penetrated more than half the buffer) jobs. This should be kept on a daily basis and posted on a weekly basis. By making the areas of potential constraint more visible, you will increase the urgency of solving the problems that create those "almost late" orders. Some people will not like it, because their area is being shown as the most problematic. Good! They will be more motivated to move work faster to the constraint, thus achieving the purpose you want — faster throughput. You will have the opportunity to see where potential bottlenecks exist, before they affect the customer, buying you time to find a solution.

In Figure 7.20.2, the procedure is developed for a functional organization. If you manage using a focused-factory concept, the assignments will be different, but the roles will be the same. The buffer management procedure gives immediacy to the schedule. If you miss the schedule, you lose throughput. So, you meet every day to ensure that everyone is on the same page and shares your concern for the health of the company.

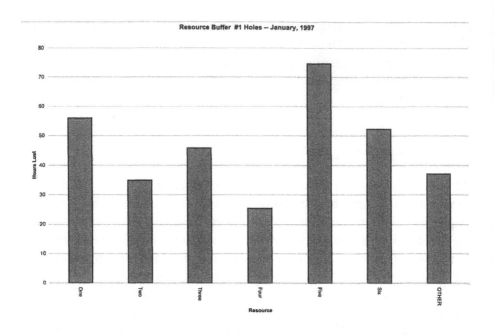

Figure 7.20.1a   Buffer Penetrations

Resource Buffer #1

| Date | Resource # | | | | | | | |
	One	Two	Three	Four	Five	Six	OTHER	Total
1/13								
1/14	10	7	9	7				23
1/15	9	8	4	14		10		35
1/16	5	1	4	4		15		24
1/17	3	14		7	4	4	10	38
1/20	11	6				10	1	17
1/21	9		16			3	12	31
1/22	1				1		15	15
1/23	1				13			13
1/24					9	0		9
1/27			10		9	11		30
1/28			15	10	15			40
1/29								0
1/30								0
1/31								0
TOTAL	48	36	58	41	51	52	38	
FREQUENCY	8	5	6	5	6	7	4	

Figure 7.20.1b   Buffer Penetrations Table

Purpose

Analyze and control the actual and planned content of the buffers, correctly allocate resources, maximize throughput, minimize response time, and maximize on-time delivery performance.

Resources

1. Drum schedules for the current day, plus the next 1 or 2 days, depending on the time of day. (Current day plus the next day in the AM, and current day plus two days in the PM.) These schedules represent the planned resource buffers.

2. Scheduled shipping report for current day plus the next day for all products. This report represents the planned shipping buffers.

3. Identification of all jobs missing from the resource and shipping buffers. This information is provided in the Buffer Management reports for both resource and shipping buffers.

Accountabilities

The planner is accountable for the status of the shipping buffer. The buffer is set according to the buffer policy. Completed jobs should arrive at shipping at least half the buffer before the scheduled ship date.

The material handling department supervisor is accountable for the status of the constraint resource buffer. The resource buffer is set according to the buffer policy. All jobs passing through the constraint (drum) resource should be in the constraint queue at least half the buffer prior to the scheduled drum date.

Supervisors and/or assigned buffer managers are accountable for keeping the manufacturing manager and scheduling advised as to the status of the buffers and buffer shortages on a daily basis, through normal daily communication and the daily buffer meetings.

Buffer managers may, as buffer shortages dictate, reorganize jobs in nonconstraint resource queues to ensure that buffer shortages are worked on first or next up. Buffer managers are accountable to the manufacturing manager not only to identify the shortages in the buffers, but also to act to get the shortages to the buffers *immediately.*

The manufacturing manager is accountable to execute the manpower planning policy and procedure to eliminate capacity shortfalls at nonconstraint resources in order to preserve the buffers.

Procedure

1. Hold meetings daily at the beginning of first shift — preferably as huddles, rather than sit-downs. Duration should be no longer than 15 minutes.

2. Review current day buffers and missing tasks.

3. Verify status of material, supplies, and personnel resources to execute the schedules.

4. Verify working order status of any equipment resources.

5. Identify locations of jobs missing from the buffers.

6. Define action plan and assign accountabilities for resolution.

7. Record results in the buffer management spreadsheet.

8. Review the Buffer Management report weekly for the top three resources causing holes in the buffers.

Figure 7.20.2 Buffer Management Procedure

<u>Purpose</u>

Communicate the global manufacturing strategy to other functions within organization.

Provide a formal opportunity to identify potential problems with the strategy.

Create consensus to execute strategy.

<u>Process</u>

1. Master schedule prepared, reconciling sales forecast to manufacturing capacity.

2. Meeting convened to review schedule with the following functions represented: manufacturing, engineering, customer service, sales, and scheduling.

3. Schedule presented by scheduler.

4. Comments from attendees.

5. Action items reviewed and identified.

6. Revisions and approvals to schedule.

<u>Comments</u>

Review meeting held monthly (minimum).

Figure 7.21 Master Schedule Review Process

In-Depth TOC Practitioner Training

Long term, your success depends on your ability to create and develop people skilled in managing the system. That means they will have to understand constraint management concepts. The key to a lasting implementation is to form a deep understanding of CM principles within the organization. I have to admit that, at best, I was only able to effectively train about three people per company. I think that is because there are not many people willing to make the investment to learn. There are just too many pressures to attend to in running the business.

Nevertheless, be prepared to create experts. They will not be cheap. You will have to develop more than one. You may become the expert yourself. Drum–buffer–rope is only the tip of the iceberg. I have been implementing for over 10 years, and I think I still have more to learn.

Master Scheduling Process

This process is designed to communicate the medium- to long-term global manufacturing strategy to other functions within organization, provide a formal opportunity to identify potential problems with the strategy, and create consensus to execute the strategy (Figures 7.21, 7.21.1, and 7.21.2). It is a way to synchronize the long-term plan with the short-term plan. In a

Week of	Planned Production Rate	Prerelease	Released	Current Schedule
31-Jan	$750,000	$0	$525,221	$525,221
7-Feb	$750,000	$946	$872,528	$873,474
14-Feb	$800,000	$21,781	$669,423	$691,204
21-Feb	$850,000	$42,463	$854,004	$896,467
28-Feb	$900,000	$463,878	$587,803	$1,051,681
7-Mar	$950,000	$698,005	$0	$698,005
14-Mar	$1,000,000	$1,210,022	$96,700	$1,306,722
21-Mar	$1,000,000	$1,014,799	$0	$1,014,799
28-Mar	$1,000,000	$683,048	$60,787	$743,835
4-April	$1,200,000	$0	$0	$0
11-Apr	$1,200,000	$117,774	$0	$117,774
18-Apr	$1,200,000	$96,534	$0	$96,534

Note: Rates are in dollars because that is the unit of measure for capacity at the plant at which this was developed. You can use hours, pieces, or any other unit that is meaningful to you.

Figure 7.21.1   Master Schedule Review Worksheet

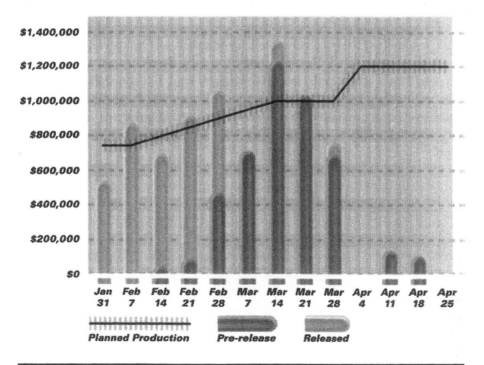

Figure 7.21.2   Master Schedule Review Chart

seasonal business, it is critical to implement the master scheduling process to get control of the business. Sometimes, this is called "sales and operations planning." Entire books have been written about how to do it correctly. The most important requirement is to base the master production schedule (MPS) on the demonstrated capacity of the constraint. The MPS review process allows management to make informed decisions about inventory investment, capacity, and service levels far ahead of when the order is actually received.

Summary

You now have all the examples you were looking for in Chapters 1 through 6. I cannot stress enough the importance of documenting your new processes and policies. The documentation is important for learning — both for the implementer and the organization. The implementer gains understanding through verbalizing the needs of the organization, and the organization gains in the knowledge that is being shared. With documented policies and procedures, you teach, communicate, and learn.

If you are successful implementing all the procedures and policies outlined here, you will probably be promoted. Your results will be spectacular. Do not let it go to your head. Getting control is only foundational to long-term success. This is where most implementations stop, i.e., scheduling the plant. The real bottom line results come when you start attacking the market with your newfound capability. That is the purpose of aligning the organization-to give it the ability to understand and use its capabilities most effectively.

8 Organizational Alignment

ligning the internal functions of your business to the constraint —
both the current and strategic — anchors your implementation and
promotes continuous improvement. It allows you to focus the nonproduction resources to maximize your return on investment — systematically.

Have you ever been involved in a sales campaign that went wrong? You
sold more product, but the profits did not increase as expected? Or the demand
was so great that the plant was unable to respond properly? These are symp-
toms of misalignment between the operations and sales/marketing organizations. This seems to be the most common misalignment — the sales and
marketing strategy/tactics are not in line with operations reality or strategy
(and vice versa).

Organizational alignment is when "all elements of a company work
together *in concert* within the context of the organization's core ideology and
type of progress it aims to achieve — its vision or goal" (Collins and Porras,
p. 90). "The effect of alignment is that people receive a consistent set of signals
to reinforce behavior that supports the core ideology and achieves desired
progress" (p. 215). Thus the goal of organizational alignment is to have everyone on the same page, moving toward the goals the organization has established for itself. In TOC terms, everyone is making decisions that will increase
throughput, reduce operating expense, and reduce inventory.

The issue of alignment is an extension of step 3 of the focusing steps,
"Subordinate Everything Else to the Above Decisions." To attack the problem,
we can break it into related questions:

- What decisions have been made already (to exploit the constraint)?
- Where is the constraint today?
- Where does management want it to be?
- Where is it likely to move next?
- Is that acceptable?
- What do I do right now?

Therefore, to achieve proper subordination (alignment), you must know where the organization is relative to the desired constraint. You must also know how your department/team affects the constraint and the suborganization's plan relative to it. Finally, you must know how your individual job affects the constraint and what decisions can improve the exploitation of it.

Creating alignment in the organization allows you to systematically address the fourth step of TOC, "Elevate the System's Constraint." The question is: How do you do it?

The problem of alignment can be broken down into three segments: strategy development, measurement, and response. Therefore, your implementation plan focuses on developing a medium- to long-range plan for the constraint resource, establishing measurements of the inputs and outputs of the constraint, and ensuring the organization will respond to the measurements.

Strategy Development

Your business strategy *must* consider the constraint of the business. You already know that or you would not have gotten this far. But, to develop a workable strategy requires thinking about where the constraint is today and predicting its movement to different locations at some time in the future. Probably, you do not know how the constraint will move. Two tools will enable you to roughly predict where the constraint will move based on an assumed sales forecast. One is the throughput contribution by product report; the other is your buffer management report.

Throughput Contribution

You increase profits by increasing the rate your business satisfies customers. Since the constraint limits your output, increasing the rate the constraint delivers customer satisfaction will increase overall output. However, the constraint will only do the work it receives. Therefore, if the work you accept yields higher dollars per day, hour, or week, then you increase overall revenue per day, hour, or week.

Put another way, if you think of your business as an hourglass, the total amount of sand you produce is limited by the size of the bottleneck. You can increase the amount of sand that flows in a period by making the bottleneck larger. However, not all the sand flowing through the bottleneck is the same. Some of that sand is valuable gold. Some of it is silver. Some of it is just plain sand. If you know which is which, you can modify the mix to increase the amount of gold and silver and decrease the amount of just plain sand. This is the goal of the product marketing and sales strategy — to change the makeup of the sand by shifting the ratios to higher content gold and silver, thus increasing the dollars flowing through that bottleneck.

In addition, if your constraint resource (if internal) is underutilized, turning work away is *very* poor subordination to the constraint! *Any* work you get will be better than having none, regardless of your costs![12] So another factor must be considered in sales planning — the capacity situation of your plant. If it is underloaded, then increased sales efforts, selective price reductions, or promotions to stimulate demand are indicated. If it is full, or nearly full, the next order you accept may prevent you from accepting another. Therefore, you must be selective about the work you accept in order to avoid filling the plant with just plain sand and blocking the gold and silver.

The purpose of the throughput contribution report is to identify those products that generate the most throughput for the least constraint contribution — the gold and silver. This information can then be evaluated in light of the current sales and marketing strategies. You can then stimulate demand on those products that deliver the most contribution per hour or depress demand on those that deliver the least contribution per hour.

To highlight the difference between product priorities set by the traditional costing approach and the TOC approach, see Figures 8.1 and 8.2. Based on the margin analysis, the most profitable products are thingamabob B, gizmo A, and then gizmo B. However, those products that generate the most throughput per hour are whatzit B, thingamabob A, and then thingamabob B. Thus, with this information, your product sales priorities (and emphasis) will change.

Based on the results of the analysis, you will know which products to emphasize and which to deemphasize. The easiest method of stimulating and depressing demand is by manipulating sales price. But, since elasticity of demand is generally not specifically known, changes in price must be done

---

[12] Assuming the sales price exceeds the total variable costs associated with that order.

Product	Sales Price	Mat'l Cost	Direct Labor	Overhead	Total	Margin	Margin %	Rank
Widget A	$88.00	$62.00	$3.79	$9.07	$74.86	$13.14	14.93	6
Widget B	$95.00	$62.00	$5.30	$12.69	$79.99	$15.01	15.80	4
Whatzit A	$195.00	$140.00	$7.58	$18.14	$165.72	$29.28	15.02	5
Whatzit B	$210.00	$166.00	$3.79	$9.07	$178.86	$31.14	14.83	7
Gizmo A	$42.00	$20.00	$3.79	$9.07	$32.86	$9.14	21.76	2
Gizmo B	$44.00	$22.00	$3.79	$9.07	$34.86	$9.14	20.77	3
Thingamabob A	$230.00	$174.00	$6.82	$16.32	$197.14	$32.86	14.29	8
Thingamabob B	$300.00	$174.00	$17.05	$40.80	$231.85	$68.15	22.72	1

Figure 8.1 Product Margin Analysis

Product	Price	Mat'l Cost	Thruput	Time per Part	Thruput/TPP	Rank
Widget A	$88.00	$62.00	$26.00	10	$2.60	5
Widget B	$95.00	$62.00	$33.00	14	$2.36	6
Whatzit A	$195.00	$140.00	$55.00	20	$2.75	4
Whatzit B	$210.00	$166.00	$44.00	10	$4.40	1
Gizmo A	$42.00	$20.00	$22.00	10	$2.20	7
Gizmo B	$44.00	$22.00	$22.00	10	$2.20	8
Thingamabob A	$230.00	$174.00	$56.00	18	$3.11	2
Thingamabob B	$300.00	$174.00	$126.00	45	$2.80	3

Figure 8.2 Throughput Contribution Analysis

selectively and conservatively. (You can usually test price elasticity by doing a promotion.)

Pricing Policy

The most prevalent approach to setting and evaluating price uses internal cost plus a fixed margin percentage for profit. This approach leaves a significant amount of unrealized profit. Costing theory states that, "Our costs will generate prices that are in line with the market's expectations." In other words, we will use a known data point (cost) and extrapolate a price from it. This approach consistently delivers incorrect product sales priorities. Consequently, many manufacturers are discarding it (Johnson and Kaplan).

The reality is that price is a function of your customer's perception of availability and value. Price is not a function of your costs. Therefore, those companies that attempt to sell on "cost" are doing no better than selling on "price" (rather than value). Although it is generally recognized that the current pricing models are imperfect, we continue to use them because no acceptable substitute is widely available.

Regardless, the question for the sales manager remains: "How do I know if I am making money on *this* order (or product)?" You make money on every order where the sales price exceeds totally variable costs, but how much? Ask a different question. Is this price the *best* price for this particular job?

Best Price

The best price is one that ensures you get the order you want with revenue better than most of the other orders you are currently receiving. Thus, the two essential questions every manufacturer must answer to maximize the pricing aspect of his order acquisition process are, is this the right order for me? And, is the revenue on this job better than I could normally receive?

Best Orders

What is the *best* type of order (job)? It is the one that most leverages *your* capital assets, leads *your* company to where you want to be, and plays to the strengths of *your* firm. A *good* pricing process allows management to determine the kind of work that is best suited for your business, then establishes pricing policies that encourage well-suited work and discourages ill-suited work. This will allow you to leverage your existing investments in the business.

Therefore, the pricing puzzle is best viewed from the perspective that asks, "How can I encourage my customers to buy products that will maximize my return on investment" (those that have the highest T/CU ratio)? The orders that are best for your company are those that have the highest T/CU ratios.[13]

Capacity Considerations

The pricing policy should be flexible enough to deal with changing capacity. When the plant is underloaded, pricing should decrease to stimulate demand. When the plant is overloaded, prices should increase to depress demand. Those products not consuming constraint resources are ripe for promotion. Those products that consume constraint resources must be carefully monitored to maintain a consistent sales level.

In situations where the plan for the constraint is to keep it in the market (growth strategy), your plan will not be centered on pricing policy to manipulate the product mix, but on capacity changes to bring the T/CU ratios to a specified (planned) level.

The pricing policy is only one part of the business plan, but it is a significant part. Throughput per hour will consistently give you accurate product priorities. From there, you can build a sales or market plan aligned with your constraint.

Buffer Management

Predicting the location of the constraint in the future is essential to building a robust plan. If you have taken care in the early stages of the implementation, you will have already decided where the constraint *should* be. Keeping it there is another matter. As the product mix and the reality on the shop floor changes, the constraint/bottleneck will want to move. Your objective in developing the plan is to keep the constraint where you want it.

Buffer management report history tells you where the constraint is likely to emerge. In Figure 8.3, you will see that Resource 6 has experienced the most disruptions to the buffer. This, then, would be the most likely candidate for the next constraint. Any efforts to elevate the constraint must consider this. Your strategy to increase capacity at the stated drum would also include a capacity increase at Resource 6 (and possibly 5 and 3).

---

[13] Assuming they also fit within the strategic framework of the organization.

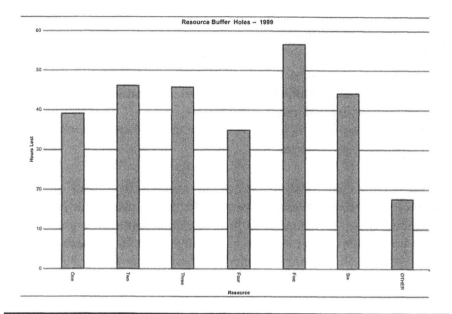

Figure 8.3   Buffer Penetration Analysis

Measuring Performance

Behavior follows measurement. This assumption is fundamental to creating alignment in the organization. If we measure the organization appropriately, we will modify the behavior in the desired way. Therefore, the problem of creating alignment through measurement and subsequent response to the measurement is in understanding the *right* things (behaviors) to measure.

Departmental (function) performance occurs at two levels: the extent to which each function performs effectively in terms of its own productive output and the extent to which it contributes to the effectiveness of the larger business unit of which it is a part. Thus, the criterion for evaluating the effectiveness of an organization incorporates the effectiveness of the functions that constitute it and the effectiveness of the business unit as a whole (a local performance measure integrated into a global performance measure).

The effectiveness of a particular business-unit is dependent not only on the effectiveness of each subordinate function (exploitation), but also on how well the work of the performing functions are integrated (subordination). It depends not only on the efficient application and effective utilization of resources within a function (constraint utilization), but also on efficient distribution of resources among functions (ensuring nonconstraints do not become overloaded).

The measurements, therefore, are structured with the goal of creating a system that will deliver three specific outcomes:

1. People pay attention to what customers want and deliver it.
2. People work continuously to improve the business processes that deliver valued customer outcomes.
3. The organization is developed to encourage cross-functional flexibility and responsiveness.

The measurements in Figure 8.4 are not meant to be inclusive, but to be examples of the kinds of measures you need to have in place. I have attempted to measure each of the elements of the order fulfillment process to accurately diagnose problems within it. Select and implement the measurements as needed. Implement too many at once, and you will have more information than you can effectively absorb and respond to.

Culture Issues

I think the most difficult problem of alignment is getting the organization to understand and respond to the new measurements. In some cases, these will be replacements to existing measures. In others, they will be brand new. In still others, the whole idea of measurement is a new thing. In each case, there are significant obstacles to implementation of a new measurement system. The problem can be distilled into three questions:

> Is the measurement relevant?
> Does the measurement accurately reflect performance?
> What do I do about it?

The answer to each of the three is training — training on CM principles, training on the measurements themselves, and training how to respond. That is why you have to have some experts in the company. If you do not have people within the organization who can articulate the measurements and their impact on the global measurements, you do not stand a chance of getting them implemented. Remember than when you first implement these measurements, they probably will not be an accurate reflection of performance. Your data has to be cleaned. Set expectations correctly.

Report	Description	Reporting Medium	Prepared By	Posting Location
	SALES Mission: Generate more profitable sales			
Throughput booked	Throughput dollars booked in period	Line graph in $	Controller	Sales
Sales effectiveness	Throughput dollars booked divided by constraint resource hours booked	Line graph in $	Controller	Sales
Sales productivity	Throughput $ booked ÷ sales expense $	Line graph in ratio	Controller	Sales
Sales $ bookings	Total $ value orders booked	Line graph in $	Controller	Sales
Quotations sales $	Total $ value orders quoted	Line graph in $	Sales manager	Sales
Quotation throughput $	Total throughput $ value orders quoted	Line graph in $	Sales manager	Sales
Quote to close	Ratio of orders quoted to orders received	Line graph in ratio	Sales manager	Sales
Quotes outstanding	$ value of unresolved quotations	Dollars — Line graph in $		
	PLANNING AND SCHEDULING Mission: Ship orders on time			
Percent on-time deliveries — dollars	Order $ shipped during the month divided by order dollars shipped on or before due date	Line graph in %	Materials manager	Materials
Percent on-time deliveries — orders	Number of orders shipped during the month divided by number of orders shipped on or before due date	Line graph in $	Materials manager	Materials
Service level replacement parts	Number of line items (of replacement parts) shipped within 24 hrs of order entry ÷ total line items (of replacement parts) shipped	Line graph in %	Materials manager	
	MANUFACTURING Mission: Produce more product while consuming fewer resources			
Orders Shipped — $	Total $ orders shipped	Line graph in $	Controller	Mfg.
Throughput $ shipped	$ shipped less totally variable cost (TVC = raw materials and subcontractors) for those orders	Line graph in $	Controller	Mfg.

Continued

Figure 8.4 Monthly Operating Measurements

Report	Description	Reporting Medium	Prepared By	Posting Location
	MANUFACTURING (*Continued*)			
	Mission: Produce more product while consuming fewer resources			
Response time	Receipt of order (date) to ship (date)	Line graph	Plant manager	Mfg.
Constraint utilization	Actual hours run ÷ hours available at constraint resource	Line graph	Controller	Mfg.
Constraint rework hours	Quantity constraint hours spent on rework	Line graph	Manufacturing manager	Mfg.
Constraint productivity	Hours earned at the constraint resource ÷ hours spent at the constraint resource	Line graph in %	Manufacturing manager	Mfg.
Constraint schedule attainment	Part hours produced ÷ part hours scheduled	Line graph in %	Manufacturing manager	Mfg.
Plant productivity	Throughput dollars shipped ÷ by production expenses incurred during period	Line graph in ratio	Controller	
Manufacturing effectiveness	Throughput dollars shipped ÷ constraint hours consumed	Line graph in ratio	Manufacturing manager	
Scrap $	Total $ scrapped during the period	Line graph in $	Manufacturing manager	
Rework hours	Total hours spent on rework during the period	Line graph	Manufacturing manager	
Production lead time	Throughput $ ÷ WIP inventory (at purchase value) at period end	Line graph in days	Manufacturing manager	
	PRODUCT (INVENTORY) PLANNING			
	Mission: Minimize investment in parts inventory			
Inventory value	$ value of inventory broken down by raw, WIP, and finished goods	Line graph in $	Materials manager	
Inventory turns — raw material	($ rm inventory consumed this month ÷ $ value of rm inventory on hand) × 12	Line graph in ratio	Materials manager	
Inventory turns — finished goods	($ fg inventory consumed this month ÷ $ value of fg inventory on hand) × 12	Line graph in ratio	Materials manager	
Material content	Ratio of rm $ to sales dollars	Line graph in ratio	Materials manager	

Figure 8.4 *Continued*

Report	Description	Reporting Medium	Prepared By	Posting Location
		GLOBAL MEASUREMENTS		
$ backlog	Total $ of open orders booked, not shipped	Line graph in $		
Return on sales	Net operating profit ÷ net sales	Line graph in ratio		
Return on net assets	Net operating profit ÷ net assets	Line graph in ratio		

Figure 8.4 *Continued*

Summary

I have focused quite a bit on the alignment of sales with operations; that is where the opportunity is. But other areas you should think about are product engineering, information technology, human resources, and purchasing/supply chain management. There are opportunities in outsourcing and in developing the make/buy decision support system. In product engineering, how much design thought is given to ensure the constraint is not affected (or that effects are minimized)? Information technology has been focused on automating the scheduling and order fulfillment processes. Very little has been done in the area of decision support.

That being said, there is enough here to last you for a year or two. The same principles apply. The same questions have to be answered:

- What decisions have been made already (to exploit the constraint)?
- Where is the constraint today?
- Where does management want it to be?
- Where is it likely to move next?
- Is that acceptable?
- What do I do right now?

You cannot quantitatively answer these questions unless you have the numbers — and the numbers are correct. Measure performance and assume the role of a coach to teach the team what they mean and how to respond.

9 Last Thoughts and Ruminations

s I have written this, I realized that I have not talked much about failure. What could go wrong? Are there any risks to the implementation? Success is a lasting implementation that results in a process of ongoing improvement. I have been involved in many implementations that produced favorable results, but there were some undesirable effects:

- Implementation of philosophy has not lasted
- Results improved, then got worse
- Consultant was fired (me)
- Implementation was canceled
- Change agent was fired
- It took a long time to see the results
- The implementation never got off the ground
- There was only localized improvement

The single largest contributor to a successful implementation is the champion of the project. When he or she believes in and is involved in the details of the project, it is successful. When the champion is the CEO and he or she leaves the details to the staff, the success has been less consistent. When I was involved in a project with people who wanted the project to be successful, it usually was. When they did not have the vision to implement, the project failed. Every great implementation has been led internally by a smart, committed, change agent. In situations where the senior executive was committed, but the people were not, the project failed.

Sometimes, people are incompetent. I have seen that. A friend of mine played football in the NFL. He told me once that "great players make great plays." The corporate world is full of people who are not suited for the jobs they do every day. They do not do them well. Do not pick these people to run your implementation. Pick your star.

Do not depend on your consultant to lead the implementation, unless you want him or her to live in your business for a while. This will be very costly and the results will come slowly. A better reason to not have the consultant run your implementation is that he or she does not understand your business. When I say understand your business, I mean the details of how your system works and the nuances of culture and customer requirements. Many, many policies and systems are informal. If you have the consultant run the implementation, these nuances will be missed. You will have a high risk of making a bigger mistake, causing more damage than the implementation is intended to cure (I know from firsthand experience!).

I have seen some very good local implementations. By local, I mean scheduling only or in the plant only. The problem with these kinds of implementations is that, if the support from management evaporates, so does the project.

Changes in management have killed implementations. The new person does not understand the system, and returns to the conventional way. I think I will leave that particular problem alone. It is how the world works; I will leave the problem for someone else to answer. But, do not let that be an excuse for not doing anything. Having good results for a time is far better than no improvement at all.

Sometimes, the organization is not ready to change. If you want to jeopardize your job, begin the implementation before you get buy-in from the organization. If you are the senior manager, without the buy-in from your subordinates and another champion, you will not be able to get the organization to move. That is why one of the first things you do is education.

It is easy to get distracted by the fires you fight every day. If you do, the project will move slowly. You might get into the project and discover you have a bigger problem than synchronizing the business. Product quality problems. Competency problems. Your supply chain is disrupted. Cash shortages. All along, I have been assuming your business is healthy in these areas. If it is not, do not begin your implementation. There is a big difference between satisfying the "necessary conditions" and making progress toward the goal. Before you optimize engine performance, make sure you have good tires. As an outsider, I have not always been able to tell when the necessary conditions are being met and when my clients needed to improve. So, I had to learn. The hard way.

The implementation process and system designs I have outlined work. Launch the project, using concrete, measurable objectives and get buy-in to implement. Do a thorough assessment and map the processes as they are today. Map out your new system design. Do a simulation if you can. Make sure your production and material planning processes are robust. Implement a measurement system to align the organization.

This book is by no means a complete guide to implementing constraint management. I think, that when you have done your implementation, you will see how far you have to go. I am now realizing how little I know. If that is you when you are done, you will have a successful implementation. Because success is not measured by the results you achieve, but by the journey you are on.

Good Luck and Godspeed.

References

Collins, James C. and Jerry I. Porras, *Built to Last: Successful Habits of Visionary Companies*, HarperCollins, New York, 1994.

Gerber, Michael E., *The E Myth — Why Most Small Businesses Don't Work and What to Do about It*, HarperCollins, New York, 1995.

Goldratt, Eliyahu, *Critical Chain*, North-River Press, Croton-on-Hudson, NY, 1997.

Goldratt, Eliyahu, *It's Not Luck*, North-River Press, Croton-on-Hudson, NY, 1994.

Goldratt, Eliyahu, *The Haystack Syndrome: Sifting Information out of the Data Ocean*, North-River Press, Croton-on-Hudson, NY, 1990.

Goldratt, Eliyahu M., *Theory of Constraints*, North-River Press, Croton-on-Hudson, NY, 1990.

Goldratt, Eliyahu and Jeff Cox, *The Goal*, 2nd ed., North-River Press, Croton-on-Hudson, NY, 1992.

Goldratt, Eliyahu M. and Robert E. Fox, *The Race*, North-River Press, Croton-on-Hudson, NY, 1986.

Johnson, H. Thomas and Robert S. Kaplan, *Relevance Lost: The Rise and Fall of Management Accounting*, Harvard Business School Press, Cambridge, MA, 1987.

Schragenheim, Eli, *Management Dilemmas: The Theory of Constraints Approach to Problem Identification and Solutions*, St-Lucie Press, Boca Raton, FL, 1999.

Senge, Peter M., *The 5th Discipline: The Art and Practice of the Learning Organization*, Doubleday Books, Garden City, NY, 1990.

Umble, M. Michael and Mokshagundam L. Srikanth, *Synchronous Management: Profit Based Management for the 21st Century*, Spectrum Publishing Company, Gilford, CT, 1997.

Reference

Index

Milton Keynes UK
Ingram Content Group UK Ltd.
UKHW031134141024
449569UK00006B/193